Creating Experience-Driven Organizational Culture

Creating Experience-Driven Organizational Culture

How to Drive Transformative Change with Project and Portfolio Management

Al Zeitoun
Bethesda
Maryland, USA

Published by John Wiley & Sons, Inc., Hoboken, New Jersey.
Published simultaneously in Canada.

Library of Congress Cataloging-in-Publication Data applied for:
Hardback ISBN: 9781394257010

Cover Design: Wiley
Cover Image: © ai-generiert-mann-technologie

Set in 9.5/12.5pt STIXTwoText by Straive, Chennai, India

SKY10089893_110424

To

the memory of
my dad
for instilling in me the passion for cultures and systems
may his leadership legacy continue to inspire generations to come.

Dad, this book is for you as I know you would have loved to write it.

Contents

Preface

Academia and researchers have written about corporate cultures for decades. What we are now discovering, as you will read in this book, are the impacts that various activities can have on cultures. There are several types and definitions of a corporate culture based upon the products offered, markets served, and core values established by senior management. As a simple definition, a corporate culture is the set of beliefs and behaviors established usually by senior management as to how employees should interact during the execution of business transactions.

Cultures normally include combinations of management's expectations, established business core values, and ethical considerations. Good cultures that are supported by the employees increase productivity and collaboration among the employees and provide them with career advancement opportunities.

The tools and techniques that program and project managers are expected to use could have a positive impact on the culture and possibly cause transformational changes. Examples include Agile, Scrum, hybrid methodologies, and the growth of digitalization. It is still uncertain the impact that artificial intelligence practices will have on the corporate culture.

For years, executives established a corporate culture and allowed each project, and even each functional unit, to establish its own culture knowing it might conflict with the corporate culture. When discussing projects, this was allowed knowing that the projects would eventually come to an end and the project-conflicting cultures would disappear. Today, companies are establishing one corporate culture that also satisfies most of the projects as well.

As discussed in this book, activities involving innovation can seriously impact cultures. Establishing one corporate culture is usually easier if the organization supports continuous rather than part-time innovation that focuses on process improvements, better materials, higher levels of quality, lower manufacturing costs, and faster time-to-market. The downside risk with continuous innovation is that workers may ignore business as usual activities.

Despite good intentions by senior management, there are always certain types of projects that can create issues with cultures. Projects involving innovation and creativity are such examples. Let us consider an innovation project that has four phases:

- Innovation project ideas
- Innovation project selection
- Innovation project execution
- Innovation project commercialization

Virtually, all companies encourage their employees to identify innovation opportunities. Companies like Google and 3M ask employees to spend a certain amount of time each day or week to work on new innovative ideas and have established policies whereby 20–30% of each unit's revenue each year must come from new products introduced within the past four to five years. Failure to do so could impact salaries and bonuses. Some employees may resist this type of innovation pressure that might remove them from their comfort zones. There is also the risk of distracting workers from core tasks.

Selecting innovation ideas for execution is usually performed by senior management. Unfortunately, their decision may be to provide most of their support for ideas that are applicable to ongoing businesses that can favorably impact short-term cash flow and profitability rather than new businesses based upon radical innovation practices that are accompanied by financial uncertainties. This can protect executive salaries, bonuses, and retirement plans but could impact long-term success.

Executives may also select only those potential projects that fit the company's core values. The core values can include target markets, age and sex of the consumers, product selling price, and marketing and sales force experience with the possible new product.

The execution of innovation projects, especially those involving radical innovation, can take quite a bit of time because of the risks and uncertainties. Employees may resist working on some radical innovation projects if they fear that they may not be able to return to their previous functional position at project completion. There may also be a fear of losing career advancement opportunities.

There is a growing trend in companies that radical innovation project teams should remain together for the commercialization of the project. This can create cultural disagreements with existing production organizations if new equipment and facilities must be purchased, new suppliers must be brought on board, and new quality standards must be met.

It is almost a certainty that innovation project management practices and the other activities discussed in this book will have an impact on cultures. The question, of course, is how much of an impact?

All of the topics discussed above, and more, appear in Dr. Zeitoun's book. If your organization wishes to achieve sustained transformational innovation success that other organizations have found, then this book is a **"must read."**

The future of organizational excellence may very well rest in the hands of the *inspiring future leaders*. These leaders must be able to create a culture that enables driving change, and making strategic business decisions, as well as, portfolio decisions.

2 August 2024

Dr. Harold Kerzner
Senior Executive Director for Project Management
International Institute for Learning, Inc. (IIL)

Background: What Inspired This Book

Overview

This book is inspired by the fast-changing world that has become the norm. With speed comes a responsibility to have the right scaffolding in place to ensure that the structure is supported as we go through the building process. The scaffolding in this book is the ***Culture***. When Peter Drucker stated that "Culture eats strategy for breakfast," he did not miss the mark.

Organizations continue to struggle in this decade with creating the supportive culture that makes the best use of the dynamic mix of process, people, and technology. These three classic attributes of transformation have matured over time, yet their integration, and linkages to creating the right supportive and adaptable work environment has missed the mark.

For the sake of this book, ***Experience-Driven Culture*** will be the intentional approach we choose to design an organization that creates the inspiring environment for workers, partners, customers, an extended group of stakeholders. The assumption will be that these cultures work well for organizations on the path of transforming and that execute this transformation work in the form of portfolios or projects and programs.

It is the goal of this work to be a practical guide to help learners and practitioners implement the ideas and principals shared in establishing a strong foundation for that type of culture. They are then equipped to proceed on the path of standing up a solid structure in the form of an organization and ways of working design that benefits from that strong foundation.

The Big Picture Model

As an attempt to summarize visually how the elements of this book will connect to create this culture, Figure A highlights four building blocks that are necessary to foundation of the proposed culture. This will also support the various

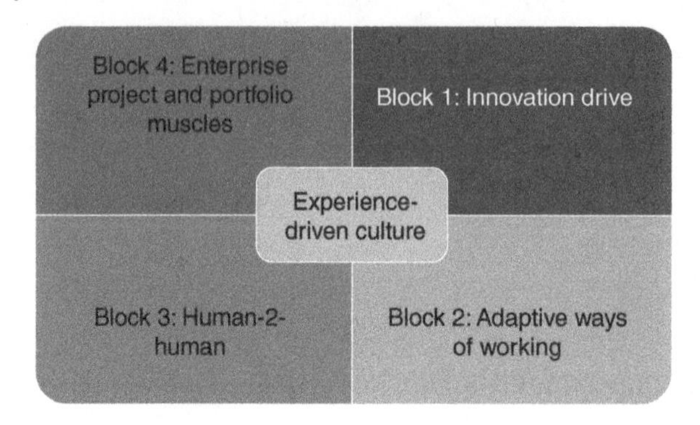

Figure A Experience-driven Culture Model.

hypotheses that will be validated with experiences, case studies, various publications, LinkedIn research, and relevant interviews.

> **Tip**
> TIP The recipe for creating an experience-driven culture requires a balanced mix of four building blocks.

The Book's Sections

The first section, **Experience-Driven Innovation with Portfolio of Projects,** will cover the importance of innovation in the world of projects and the ways for creating experiences that matter. This section will also tackle the importance of achieving integration across the execution landscape of projects within a portfolio. Most of this section elements will map to Block 1: the Innovation Drive.

Section 2, **Essential Skills to Lead Experience-Driven Cultures,** is a core section to creating these future cultures and tackles the building of the clear and visible impact muscle, the ways of creating effective experiencing, establishing the human connection, and the role of the ongoing changing and maturing digital fluency. This section crosses Block 2: Adaptive Ways of Working and Block 3: Human-2-Human.

The third Section of the book, **Creating Experience-Driven Cultures with Enterprise Portfolio Management Muscles,** addresses the critical role of portfolio management in building this Experience-Driven culture, the foundational elements of sustaining that culture, and the relevant portfolio management muscles. Most of the elements of this section map to Block 4: Enterprise Project and Portfolio Muscles.

The **Path Forward** section is focused on best practices for creating the consistency of excellence in these created cultures and then sustaining that creation.

Section I

Experience-Driven Innovation with Portfolio of Projects

Section Overview

This section describes the fundamental linkages between innovation and portfolio management. The value of portfolio management in achieving the value outcomes of organizational change is critical. A few initial tools to aid leaders in creating experience-driven innovations will be highlighted. Prioritizing with excellence is supported by a few case studies across global organizations.

Section Learnings

- The relationships among portfolios, programs, and projects to understand how they align to deliver value.
- Why do portfolios of projects enable the achievement of a competitive advantage?
- The qualities needed to develop the innovation leaders of the future.
- How the different team members' views could create the foundation for innovating with the right projects?
- Using stakeholders' alignment and innovation labs to speed innovation and build a culture that champions innovative ways of working.
- Empowering the linkages across portfolios of programs and projects with simplicity and integrating stories.

Creating Experience-Driven Organizational Culture: How to Drive Transformative Change with Project and Portfolio Management, First Edition. Al Zeitoun.

Keywords

- Strategist
- Transformation
- Ways of working
- Experience
- Ecosystem
- Value focus
- Simplicity
- Stories

Introduction

The world of work continues to mature to adapt to the demands of both scaling and sustaining future growth. The execution of work in a consistent portfolio-like manner is increasingly valuable in creating the connected rigor management teams can understand, visibly see, and empower their decision-making process. The way of leading in this dynamic environment requires a new breed of leaders capable of creating cultures that are experience-driven.

Experience-driven in the context of innovation projects, looks at the return on experience (ROE) in every project interaction. This muscle is becoming highly critical to learn and develop. This section highlights the challenging and rewarding journey of building cultures that invest in creating such a commitment. Reimagining how innovation is done in future is becoming a continuous improvement scenario. You will learn how to innovate faster and set the foundation for rewarding experiences across the portfolio of projects.

Building the necessary experience-driven capabilities requires commitment, time, and energy in exploring what it takes to adjust and build fitting ways of working. Many great global organizations have managed to get their teams moving in that direction, without even committing a name to this movement. It became natural to their leaders given the demands that were placed on their research and development units and other organizational entities dedicated to innovation.

Across the film industry, there are unique examples to showcase how a possible culture shift creates outcomes and ROE that are beyond what was envisioned. Sports and athletes also teach us about shifts created ahead of major competition or transformation projects. As the world welcomes newer generations, like Gen Alpha entering the workplace, it is going to become more of a critical business priority to ensure that we are budling those experience-rich cultures. Some of these movies and sports-related examples will be referred to in this section and other sections of the book.

There will be eight hypotheses that you will see starting in this section and will be elaborated on and tested throughout the book. They cover many of the points that are critical for creating such experience-driven cultures. They get into the areas of people, processes, technology, behaviors, values, and practices. A few of these will be repeated across the book's sections as they cross multiple learning outcomes and connect to one or more of the four building blocks of the experience-driven culture model shown earlier.

Several global case studies starting in this section will be analyzed to propel us into the design of these cultures and highlight the angles of innovating using portfolio mindset. As a future leader, you will need to fully immerse yourself in a future of continual learning in order to adjust these experiences to the needs of your critical stakeholders at that point of time and maturity. As the leader of a portfolio, you have to make it a priority to operate as a strategist. This means being capable of slowing down to go faster, which is a concept that you will find discussed in a variety of ways throughout the book.

1

Innovation in the World of Projects

This first chapter is focused on the understanding of the typical innovation in the project ecosystem. It is intended to highlight foundational innovation success principles that help the leader to design experience-driven future work environments with the diverse stakeholders' views and inputs in mind. This is intended to enable setting the stage for the ***Experience-Driven*** way of innovating to be built as a natural muscle for operating in such a project economy of today and into the future.

In a world that demands excellence in innovation and in delivering digital transformations, where digitalization continues to scale, the return on experiences (ROEs) that we create could directly contribute to the aspirational growth targets and the achievement of the most impactful missions.

Key Learnings

- Understand the value of how developing a holistic view successfully supports innovation.
- Explore a case study that looks at the critical people and behavioral shifts.
- Understanding the new strategist leader and its impact on successful transformation.
- Learn from a movie example how to develop the critical views that expedite your movement toward being an impactful, experience-driven conductor.
- Start addressing how this work's hypotheses link to driving innovation and aligning around the dynamic needs of customers and stakeholders.

1.1 The Holistic View

What makes committing to continual innovation challenging is sometimes linked to the short-term lens that organizational leaders possess. For impactful and sustainable innovation, the focus needs to be holistic in order to see the potential

Creating Experience-Driven Organizational Culture: How to Drive Transformative Change with Project and Portfolio Management, First Edition. Al Zeitoun.

over the horizon. The commitment and investment required to incubate ideas that ultimately could create the next unicorn is a difficult one. This is where being holistic comes in.

One of the critical reasons why holistic view matters is the fact that we can see beyond what is in front of us or what is obvious. It requires us to be obsessed with the problem. It is the commitment to slow down to go faster. By thinking holistically, we uncover every angle that has a strategic impact on where the innovation projects could take us and the surrounding assumptions and constraints that could hinder their success. Holistic also means that we consider the entire ecosystem that will be the birthplace for these projects.

1.1.1 Developing the Holistic View

Understanding innovation portfolio interactions complexity and finding ways to simplify it is critical. Having a holistic view has continued to be one of those most talked-about capabilities for leaders involved in managing portfolios of programs and projects. In describing an affective project manager, this quality is usually listed among the top ones. Yet, although commonly mentioned, it is seldom well practiced, nor properly invested into.

Ways to describe this capability could be big picture, end-to-end thinking, stepping outside the box, and seeing beyond the obvious. Developing the holistic view is an intentional practice. It requires us to first recognize its importance and then dedicate relentless focus to building it.

As highlighted in Figure 1.1, the abovementioned ways of describing the holistic view could be highlighted by the picture in the figure, especially the balloons

Figure 1.1 Holistic View Sample. Credit: KELLEPICS/Pixabay.

floating on the top of that busy city that seems to have been built at the foot of a mountain range or something similar. Being in the balloon allows the leader to see holistically. By stepping away from the details on the ground, the leader can see the effectiveness of the city design, or the lack hereof, and could see opportunities to innovate solutions to problems that might not have been seen before, while you are so close and are in the midst of the many busy landscapes of high-rise buildings. This capacity of stepping away and zooming out sounds simple, yet it becomes more and more challenging in today's busy and noisy work environments.

The figure also shows the importance of sharpening the focus from multiple angles. Having multiple balloons reflects viewing a given scenario from different angles and possibly creating a mix of objective ideas for the beginnings of innovation. This will remain an increasingly important power in the future of work.

1.1.2 Connecting to the Movies

Strategy for executing change is difficult. In **Remember the Titans**, Denzel Washington had to exercise being holistically strategic in building a true integration culture that was lacking, yet necessary for inspiring a joint view of success for that true story Virginia school football team.

Figure 1.2 shows the stakeholders that Denzel (Coach Boone) had to work across and align as a true holistic leader should. The figure reflects the potential complexity in maintaining the strategic clarity for the football team across such diverse body of stakeholders. Denzel had to continually adjust his lens and update what the next strategic move looked like while reflecting that in necessary big-picture directional changes.

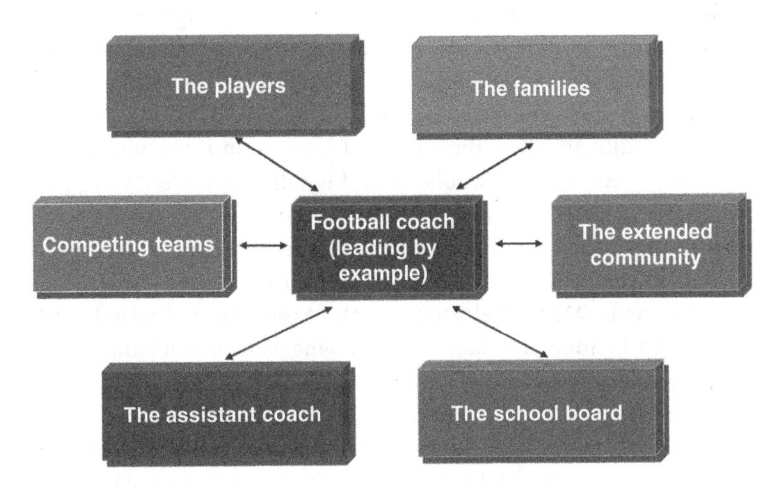

Figure 1.2 Holistic Leader's Stakeholders Strategic Links.

The sample of key stakeholders shown is all part of Coach Boone's portfolio. Handling each of these stakeholder entities was almost like an individual project for him. Building an integrated culture and overcoming racism required a natural cascade of consistent behaviors in his interactions with the players and others in the stakeholders group.

As an example, the players' grouping required him to exercise discipline and resilience on the path of building a strong team. He had to overcome personal limitations and biases and find creative ways to build critical trust and bonds among the players. This is an example of an experience-driven culture where the players had to experience what it takes to live in the world of another player of a different skin color. These rich experiences enabled the building of a well-knitted team.

Coach Boone had to maintain a holistic view that drove his season's success and supported the critical integrator role that was required to break down the team's and other stakeholders' silos.

Tip

Invest in developing a holistic view that enables the teams to experience better understanding of how their roles contribute to innovative outcomes across the portfolio.

1.2 Portfolios of Projects Matter

Although most organizations understand the importance of viewing their business in the form of a portfolio, not many excel in applying the principles of portfolio management and seeing this as a strategic competency. This assumption could limit the organizational ability for achieving the balanced use of resources across the portfolio. It could directly affect the scaling of innovation if we miss out on including the right projects in the portfolio mix. If handled positively, designing and executing against the right portfolio becomes a natural organization muscle that is to be matured over time and that brings the data, people, and technology in alignment with critical choices and their associated priorities.

In an interview with **Mary Palmieri, with Amazon, Principal PM, AI Content Thought Leader**, she shares the following views about building the portfolio management muscle:

It's almost business fundamentals you're looking at then, it's what levers can you pull to drive top line and bottom-line growth and how can you improve operations and performance to move quicker and better. I mean that's just like about iterating the view of success.

Talking about portfolio management the same way, like the Project Management Institute (PMI) away, which is a portfolio of initiatives projects, is a best practice because you do need to all be speaking the same language and getting alignment on the inputs and the outputs. For maturing Key Performance Indicators KPIs, I think it's diving deep into the use case and really getting present to what you're trying to accomplish.

On my own program, I'm owning the content acquisition program, so I've almost had to define a language for the organization on what it means to acquire content, how to measure the success of digital assets and content management, and all that stuff that's almost like it's a new language. I think there's the common business framework, but then it's like you're in your own little country, you're making up your own little tribal language that you and your team need to start speaking, and you also need to educate senior leaders on that language.

Tip

Portfolio management is a strategic muscle. Executives have to grasp its language to ensure that strategic choice-making becomes an attribute of the organizational culture.

1.3 The Speed of Innovating

Clock speed is the norm in industry today. This means that the demand for fast innovation and creation is only increasing. The good news is that computing capabilities and the advances in artificial intelligence (AI) and other simulating capabilities are all contributing to our ability to adapt facts and predict what's coming.

Time is not our friend when it comes to major climate and environmental changes, which places more pressure on the critical importance of innovation.

As in our article, Kerzner and Zeitoun (2023), even though the hot topic of AI based on large language models (LLMs), such as ChatGPT, is gaining major attention, the concept of AI itself was first proposed by British mathematician, Alan Turing, in his 1950 paper "Computing Machinery and Intelligence." We reached a state where the education of AI that we achieved, coupled with the massive amount and higher quality of data, got us to a point of impact creation on initiatives delivery that is meaningful and disruptive.

To support the future of project management and the new ways of achieving excellence, AI-enabled project work:

- Creates efficiencies that were not possible previously
- Levels the playing field where the basics of planning and executing projects can easily and quickly be covered

- Expedites the onboarding of project teams to get them to perform faster and more smoothly
- Creates a heavy focus on understanding the customer's definition of project business value
- Enhances the quality of decision-making and turns the supporting processes to a highly data-driven approach
- Opens the door for new skills and roles for the project managers and teams of the future

1.4 Managing Transformation Matters

In this section, I will use an example of a case study to highlight what an organization could experience as it goes through transformation. As in any successful transformation, it should combine a good balance between **people, process, and infrastructure/technology** to ensure a balanced and efficient progress of the transformation program toward achieving its benefits to the organization.

Case Study

Honicker Corporation[1]

Background

Honicker Corporation was well recognized as a high-quality manufacturer of dashboards for automobiles and trucks. Although it serviced mainly U.S. automotive and truck manufacturers, the opportunity to expand to a worldwide supplier was quite apparent. The company's reputation was well known worldwide, but it was plagued for years with ultraconservative senior management leadership that prevented growth into the international marketplace.

When the new management team came on board in 2009, the conservatism disappeared. Honicker was cash-rich, had large borrowing power and lines of credit with financial institutions, and received an AA-quality rating on its small amount of corporate debt. Rather than expand by building manufacturing facilities in various countries, Honicker decided to go the fast route by acquiring four companies around the world: Alpha, Beta, Gamma, and Delta companies.

1 Kerzner, 2022/John Wiley & Sons.

Each of the four acquired companies serviced mainly its own geographic area. The senior management team in each of the four companies knew the culture in their geographic area and had a good reputation with their clients and local stakeholders. The decision was made by Honicker to leave each company's senior management teams intact, provided that the necessary changes, as established by corporate, could be implemented.

Honicker wanted each company to have the manufacturing capability to supply parts to any Honicker client worldwide. But doing this was easier said than done. Honicker had an EPM methodology that worked well. Honicker understood project management and so did the majority of Honicker's clients and stakeholders in the United States. Honicker recognized that the biggest challenge would be to get all of the divisions at the same level of project management maturity and use the same corporate-wide EPM system or a modified version of it. It was expected that each of the four acquired companies might want some changes to be made.

The four acquired divisions were all at different levels of project management maturity. Alpha company did have an EPM system and believed that its approach to project management was superior to the one that Honicker was using. Beta company was just beginning to learn project management but did not have any formal EPM system, although it did have a few project management templates that were being used for status reporting to its customers. Gamma and Delta companies were clueless about project management.

To make matters worse, laws in each of the countries where the acquired companies were located created other stakeholders that had to be serviced, and all of these stakeholders were at different levels of project management maturity. In some countries, government stakeholders were actively involved because of employment procurement laws; in other countries, government stakeholders were passive participants unless health, safety, or environmental laws were broken.

It would certainly be a formidable task to develop an EPM system that would satisfy all of the newly acquired companies, their clients, and their stakeholders.

Establishing the Team

Honicker knew that there would be significant challenges in getting a project management agreement in a short amount of time. Honicker also knew that there is never an acquisition of equals; there is always a "landlord" and "tenants," and Honicker is the landlord. However, acting as a landlord and exerting influence in the process could alienate some of the acquired

(Continued)

Case Study (Continued)

companies and do more harm than good. Honicker's approach was to treat this as a project and to treat each company, along with its clients and local stakeholders, as project stakeholders. Using stakeholder relations management practices would be essential to getting an agreement on the project management approach.

Honicker requested that each company assign three people to the project management implementation team that would be headed up by Honicker personnel. The ideal team member, as suggested by Honicker, would have some knowledge and/or experience in project management and be authorized by their senior levels of management to make decisions for their company.

The representatives should also understand the stakeholder needs of their clients and local stakeholders. Honicker wanted an understanding to be reached as early as possible that each company would agree to use the methodology that was finally decided on by the team.

Senior management in each of the four companies sent a letter of understanding to Honicker, promising to assign the most qualified personnel and agreeing to use the methodology that was agreed on. Each stated that its company understood the importance of this project.

The first part of the project would be to come to an agreement on the methodology. The second part of the project would be to invite clients and stakeholders to see the methodology and provide feedback. This was essential since the clients and stakeholders would eventually be interfacing with the methodology.

Kickoff Meeting

Honicker had hoped that the team could come to an agreement on a companywide EPM system within six months. However, after the kickoff meeting was over, Honicker realized that it would probably be two years before an agreement would be reached on the EPM system. Several issues became apparent at the first meeting:

- Each company had different time requirements for the project.
- Each company saw the importance of the project differently.
- Each company had its own culture and wanted to be sure that the final design was a good fit with that culture.
- Each company saw the status and power of the project manager differently.

- Despite the letters of understanding, two of the companies, Gamma and Delta, did not understand their role and relationship with Honicker on this project.
- Alpha wanted to micromanage the project, believing that everyone should use its methodology.

Senior management at Honicker asked the Honicker representatives at the kickoff meeting to prepare a confidential memo on their opinion of the first meeting with the team.

The Honicker personnel prepared a memo including the following comments:

Not all of the representatives at the meeting openly expressed their true feelings about the project.

It was quite apparent that some of the companies would like to see the project fail.

Some of the companies were afraid that the implementation of the new EPM system would result in a shift in power and authority.

Some people were afraid that the new EPM system would show that fewer resources were needed in the functional organization, thus causing a downsizing of personnel and a reduction in bonuses that were currently based on headcount in functional groups.

Some seemed apprehensive that the implementation of the new system would cause a change in the company's culture and working relationships with their clients.

Some seemed afraid of learning a new system and being pressured into using it.

It was obvious that this would be no easy task. Honicker had to get to know all companies better and understand their needs and expectations. Honicker management had to show them that their opinions were valuable and find ways to win their support.

Review Questions

1. What are Honicker's options now?
2. What would you recommend that Honicker do first?
3. What if, after all attempts, Gamma and Delta companies refuse to come on board?

(Continued)

Case Study (Continued)

4. What if Alpha company is adamant that its approach is best and refuses to budge?
5. What if Gamma and Delta companies argue that their clients and stakeholders have not readily accepted the project management approach and wish to be left alone with regard to dealing with their clients?
6. Under what conditions would Honicker decide to back away and let each company do its own thing?
7. How easy or difficult is it to get several geographically dispersed companies to agree to the same culture and methodology?
8. If all four companies were willing to cooperate with one another, how long do you think it would take to reach an agreement and acceptance to use the new EPM system?
9. Which stakeholders may be powerful and which are not?
10. Which stakeholder(s) may have the power to kill this project?
11. What can Honicker do to win their support?
12. If Honicker cannot win their support, then how should Honicker manage the opposition?
13. What if all four companies agree to the project management methodology and then some client stakeholders show a lack of support for use of the methodology?

1.5 The New Strategist

Looking back at the Honicker case study mentioned above and reflecting on the three transformation pillars confirm that transformation matters:

1) **People:** There were gaps in alignment on the purpose of the transformation across the four companies. There were also signs of leadership weaknesses, in

addition to ego issues on the part of one of the companies that could jeopardize what is in the best interest of all companies combined. The commitment and motivation for the transformation were lacking.

2) **Process:** The change management aspects of this integration of companies around a common system and achieving a joint focus were underestimated.
3) **Infrastructure/Technology:** With the gaps in readiness across the four companies in maturity and commitment, one would assume that their further system and digital capabilities would vary tremendously. This is also coupled with the laws in the different countries that could have an impact on data regulation adherence and other legal aspects that would hinder the fast movement toward achieving the transformation outcomes.

In addition, at the center of the success of the transformation is a new form of a leader in our organizations, the ***Strategist***. With successful transformation and advances in technology and LLMs, and as AI understands the physical world better, the leader will adapt and will have more time to think again for a change and become the new strategist.

Having such a form of a leader would have tremendously contributed to the possible success of the Honicker coloration transformation. This kind of a leader would have been able to articulate a connected strategic vision that inspires all four companies' leaders and teams. The leader would have been able to anticipate the human dimension complexities better and could have positively empowered the core selected team on behalf of each company.

It is important to mention that this new strategist view of tomorrow's leaders comes with a high expectation of intense collaboration across organizational boundaries and groups of stakeholders. Similar to the classic analogy of the music conductor, as seen in Figure 1.3, the new strategist plays a role in the transformation excellence of tomorrow's organization.

In the world of transformation, this also includes the right fitting processes, as in the orchestra plans coming together, which is what a conductor needs to ensure in order to create the expected harmony, including all the prep leading to the final initiative. And then finally the technology as in the case of the conductor, as in real-world transformation, sensing how to bring the different instruments in and when, to create the magical anticipated outcomes at the other end.

1.6 Collaborating Across Diverse Views

The Honicker case used in this chapter exemplifies the diverse views of the four companies and their leaders and teams. It was likely a missed opportunity on behalf of Honicker management to make the most out of proper collaboration across these diverse views. In a highly digital universe ahead, it is more critical that

Figure 1.3 The New Strategist. Credit: chenspec/Pixabay.

the utilization of diverse views, backgrounds, and expertise are mixed together in a pot that produces the secret sauce of the experimenting cultures of the future.

This is especially more valuable since Honicker management has realized and was already looking at each one of the four companies as a project and that their task was to integrate these four projects within an integrated portfolio of these companies. To a great extent, Honicker had already realized that the project economy was upon them and that the ideal use of diverse views was to focus on the four projects, their stakeholders' alignment, and the steady movement on the strategic objectives behind the integration of the four companies.

In an era of high experimentation, the value placed on where we spend our time comes at a premium. It is no longer acceptable that we don't do proper stakeholders' analysis, planning, and clear communications strategy and plan for how we work across stakeholders' groups. The strength of a newly transformed organization lies in its ability to capitalize on all the pieces of its portfolio, especially the diverse views of the holistic sum of people's views. As was highlighted in the background section and in the overall Experience-Driven Culture Model, the Human-2-Human building block is critical in creating this necessary experience.

One of the eight hypotheses that are driving this work is: ***Leaders' top priority is adapting to customers' interests.***

To investigate this and other hypotheses, I have conducted open LinkedIn research within my network and indicated to the willing participants that their responses will provide input to this work. Most of the hypotheses received between 400 and 900 impressions. The number of votes on the driving questions was typically a small sample of about 25 votes. The intent of these questions was

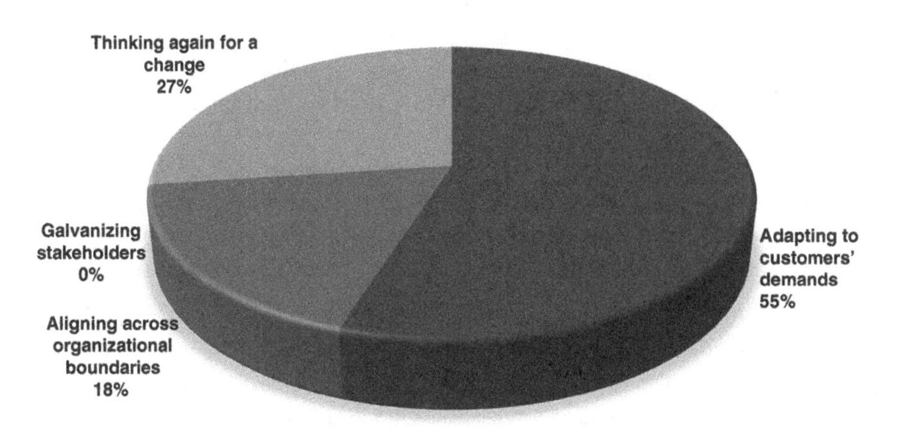

Figure 1.4 The Leaders' Priorities Hypothesis. *Note*: Based on LinkedIn Open Polling, April 2024.

to get an overall pulse of the professionals on culture, portfolio management, project management, people, digitalization, and ways of working topics.

For this hypothesis on customers, the question used was: *In your opinion, where will leaders in the future organization increasingly spend their time?*

The distribution of the answers is shown in Figure 1.4.

It is becoming increasingly valuable in this experience-driven organizational culture, the degree of focus on how to build agility and adaptability. This has been a steady muscle to grow since the time of the COVID pandemic and has only connoted to increase in importance with the number of uncertainties that surround our organizations and the demands of the markets and customers where we operate.

> **Tip**
> Adapting to customers' demands could benefit from an international collaboration across diverse views of stakeholders.

1.7 Aligning Stakeholders

As was highlighted in the Honicker Corporation case study, management and the core team that was established have not succeeded in aligning across the key stakeholders. Getting to that alignment is a core capability of successful cultures.

Honicker management could have capitalized on the early stages of this transformation initiative to ensure alignment on the reasons behind the transformation and uncover the risk appetite of each of the four companies. It is also critical to identify which stakeholder group has the highest likelihood of influencing the change and building the needed momentum to increase their interest in the transformation initiative.

In the interview with Mary Palmieri, as mentioned earlier in this chapter, she highlights multiple elements about the importance of culture in innovating and accomplishing the work of transformation initiatives. A number of the points she makes highlights the importance of aligning across stakeholders' groups:

- I do think the culture is crucial for delivering success.
- I think you always have to reward risk taking because I believe that's the difference making. I've found, in at least innovation programs, it's the willingness and the ability to take risks, and it's almost a mindset of not being afraid to fail.
- I think it's important for any company to create a safe space for innovative minds to try things and they could be calculated risks.
- I think that's the biggest challenge for sure is to get the organization there.
- I think it's about aligning to what their vision and objectives are.
- I would almost argue nowadays every company should have an innovation program like it's a mindset, it's a skill set, and it's more creative.
- I feel like there's the engineers' group, there're finance people, and others that should be like an innovation team whose job is to talk to customers, whether it's internal stakeholders or external consumers or whomever.
- The project teams should do the analysis and continuously identify problems to solve that align with the greater organizational mission.

> **Tip**
> In the right fitting culture for executing on organizations' complex initiatives, it is crucial to align across key stakeholders.

1.8 Championing Innovation

As a final key priority point in this chapter, it is critical to remember the fundamentals around the importance of the champions. Such as in the case of programs seeking to achieve value, having the right champion for innovation is critical. Innovation is a major organizational commitment. It does require us to embed creativity in what we do and how we do that. It also requires a gut check when it comes to the risk-taking muscles of the organization. A cross-executive team understanding of the delicate balance between opportunities and threats is essential, and the champion brings that clearly forward.

In the interview with **Mary Palmieri**, she continues by highlighting some important attributes for the innovative cultures of the future. Mary indicated attributes and examples that are a string direct reflection of how to thrive in building the innovation mindset and how to support and champion the consistent growth of this innovation muscle.

I think the top three attributes are:

1) Just being open-minded and having the ability to think outside the box very often. This is especially the key in larger organizations, where we're a cog in a wheel, which is great operationally, yet when you need to come up with new product solutions or program ideas, there needs to be that flexibility or innovative mindset.
2) I think the culture also needs to be data driven, so really putting the appropriate mechanisms in place to measure success and getting into the practice of defining success and working iteratively.
3) Then lastly, the talking to customers thing, just really interviewing your customers, understanding your customers, and learning the data around your customers.

Speaking of customers, you have to balance what the customer is telling you and what the problem is with what they really need. It's a chicken and an egg thing you need to achieve that product-market fit to win, yet they're going to tell you what that is.

We were pitching an AI-driven marketplace in 2016 when only 9% of the beauty industry was online, way above their heads right, but now AI is a thing people are coming around. We're the experts on that, but we had to meet them where they were, so it's like that in every situation, you almost have to meet the customer, meet the stakeholders, meet whenever where they are, and then level them up slowly but surely. Sometimes it's like little crumbs, like breadcrumbs; it's like when you're first meeting a partner, you're not going to let all the little mystery out at once!

Tip

The innovation champion plays an instrumental role in connecting the organization around what is critical to innovate: mindset, data, and customers' deep understanding.

Reference

Kerzner, H. and Zeitoun, A. (2023). Cracking the excellence code, the great project management accelerator, series article, *PM World Journal*, Vol. XII, Issue XI.

Review Questions

Parentheses () are used for Multiple Choice when one answer is correct. Brackets [] are used for Multiple Answers when many answers are correct.

1 How do you best describe the holistic view of portfolio leaders? Choose all that apply.
[ ] Able to see beyond the obvious.
[ ] Are liked by many people.
[ ] They possess big-picture understanding.
[ ] Capable of end-to-end thinking.

2 What supports new ways of achieving excellence?
( ) Tackle as many priorities as possible.
( ) Ensure tighter governance.
( ) Create a heavy focus on understanding the customer's definition of project business value.
( ) Focus mainly on implementing agile practices.

3 The Honicker Corporation management invested properly in stakeholders' management.
( ) Yes.
( ) No.

4 Which of the following provides a good analogy for the role of the new strategist leader?
( ) Being at the center of solving every portfolio issue.
( ) Ensures the team follows directions without any deviations.
( ) Creates a connected team similar to the music conductor analogy.
( ) Strong central leadership.

5 Which of the following is the building block in the experience-driven culture model that most supports the importance of stakeholders' alignment?
( ) Enterprise Project and Portfolio Muscles.
( ) Human-2-Human.

( ) Innovation Drive.
( ) Adaptive Ways of Working.

6 What are the findings that most support leaders' top priority in better connecting to what is critical for transformation success?
Choose all that apply.
[ ] Ability to say "yes" to new demands.
[ ] Ensuring that scope of transformation is frozen early in the program lifecycle.
[ ] Thinking again for a change.
[ ] Adapting to customers' demands.

7 How do you summarize key attributes of innovation champions?
Choose all that apply.
[ ] Being open-minded.
[ ] Data-driven decision-making.
[ ] They spend their time thinking.
[ ] Deep customers' understanding.

2

Creating Experiences

Leaders create experiences. These leaders are lifelong learners and value building an environment where experience adds to that learning of the teams and the organization. They know that experimentation does not hurt and that they could this way learn continuously about customers, users, partners, and other stakeholders across the ecosystem.

Creating experiences is the focus and cornerstone of the fitting future cultures in their quest for transforming and leading the digital revolution. This is what enables organizations to think, invest strategically, and create a well-connected open and sustainable future environment. In this chapter, creating experiences is tackled in multiple ways, covering the changes in how we work, how to embed experience into the DNA of the organization, build an appetite for innovation, connect across the ecosystem, and most importantly develop a culture that supports thinking and reflecting to scale the learning with speed.

Key Learnings

- Understand the changing dynamics in how we will continue to adapt our ways of working into the future.
- Learn how to develop experiences that matter.
- Explore the principles of innovation labs and how that mindset drives effective experiencing.
- Understand the possible approaches to building a connected system of people and solutions.
- Develop the thinking culture that differentiates organizations on their path to excellence.

Creating Experience-Driven Organizational Culture: How to Drive Transformative Change with Project and Portfolio Management, First Edition. Al Zeitoun.

2.1 The Changing Ways of Working

Cultures benefit from having a way of working that fits the purpose of the organization. Over the recent years, the topic of ways of working and new ways of working has dominated the business discussions and the communities of program and project managers. There is an element of this related to how and where we work, being in an office, remotely, or in hybrid mix. Then there is the element and a debate over the years across project, product, and program teams around classic ways of working versus the agile ceremonies for working, or the most commonly used hybrid of both formats.

What has become very evident is that fit matters more than anything else. In studying one of the experience culture hypotheses, my LinkedIn community was pulsed regarding the following hypothesis:

Creating ways of working that fit is a future leader's strategic focus
The question used was: *What do ways of working in the future look like?*

The results of pulsing this question confirmed the importance of choosing a fitting way of working, even at the expense of the growing interest in being a hybrid organization, or the widespread enthusiasm about the possibilities of artificial intelligence (AI) in shaping organizational design and affecting our work ways and the efficiencies of achieving outcomes. Figure 2.1 reflects these results.

> **Tip**
> In building the right supportive culture for innovation, choosing the fitting way of working ensures that the innovation teams are in their flow.

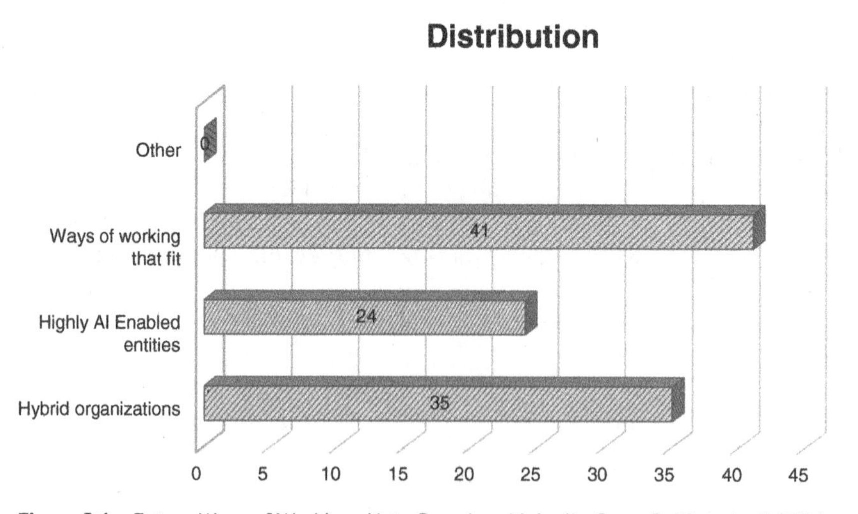

Figure 2.1 Future Ways of Working. *Note*: Based on LinkedIn Open Polling, April 2024.

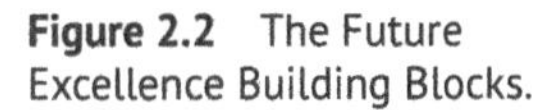
Figure 2.2 The Future Excellence Building Blocks.

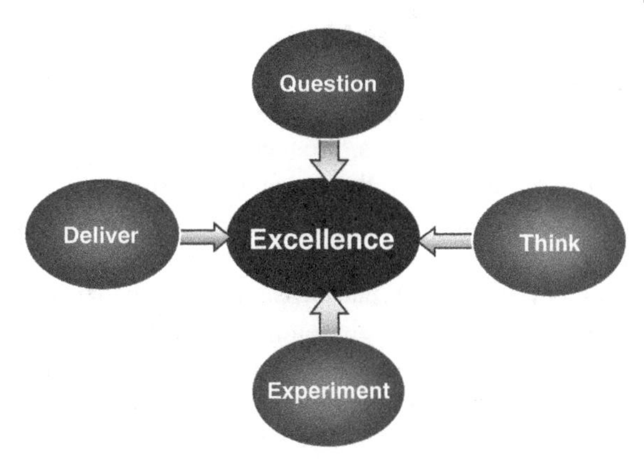

In our article, Kerzner and Zeitoun (2023), we highlight an important analogy that relates to this changing view of excellence is the focus on breaking barriers like the ones that continue to exist across organizations in this digital world.

Figure 2.2 highlights how a future excellence that is coupled with enhanced use of technology and capitalizing on the power of AI could change the dynamics of how leaders and their teams work in the future.

Program and project leaders can focus on four supporting building blocks to achieve excellence. The first block is the enhanced questioning power. The high access to quality data gives the space for practicing quality questions that are specific, have proper context, and open the door to understanding the possible complexity in a given initiative. The second building block, focusing on thinking, is one of the most valuable aspects of technological disruptions where we could have copilots and many other upcoming advances, freeing up the time that we could block on our calendars to think away from the busy project noise, and thus become more strategic.

This is also where we could then scale the amount of experimenting that the third block covers. It is about creating a knowledge-powered engine where continual learning from experimenting becomes the norm for achieving future excellence. The fourth and final building block, delivering, then starts taking on a different style as the linkages to strategy and purpose increase, while keeping the customer highly engaged in the journey to achieving aspired outcomes.

While all these blocks are important, delivery may be the most challenging. Keeping customers highly engaged may create issues if the customers are unfamiliar with the new technology being used on their projects or are resistant to the use of new concepts. Eventually, there will be alignment, but this may take time and require educating customers on new technologies and ways of working.

2.2 Experiences Matter

Saying that experiences matter is a confirmation of the human need to connect to people, things, ideas, products, and solutions. In a world where the lines of demarcation between the real and the virtual worlds have been blurred for good, it is critical that we capitalize on this digitalization and immense computation power to generate experiences that complement the human search for safe environments to play, test, and envision before finalizing an outcome of innovation and taking that into full-scale commitments. This is especially of utmost importance in a world where sustainability commitments are an increasing priority.

The following case study highlights what happens when there is no reliance on proper data to experience what a potential outcome from an infinitive could be. The case also highlights the danger when emotional decision-making affects what the project team and the market might experience as a result of such weak decision muscle.

Case Study

Irresponsible Sponsors[1]

Background

Two executives in this company each funded a "pet" project that had little chance of success. Despite repeated requests by the project managers to cancel the projects, the sponsors decided to throw away good money after bad money. The sponsors then had to find a way to prevent their embarrassment from such blunders from becoming apparent to all.

Storyline

Two vice presidents came up with ideas for pet projects and funded the projects internally using money from their functional areas. Both projects had budgets close to $2 million and schedules of approximately one year. These were somewhat high-risk projects because they both required that a similar technical breakthrough be made. There was no guarantee that the technical breakthrough could be made at all. Even if the technical breakthrough could be made, both executives estimated that the shelf life of both products would be about one year before becoming obsolete but that they could easily recover their R&D costs.

1 Kerzner, 2022/John Wiley & Sons.

These two projects were considered pet projects because they were established at the personal request of two senior managers and without any real business case. Had these projects been required to go through the formal process of portfolio selection of projects, neither would have been approved. The budgets for these projects were way out of line for the value that the company would receive, and the return on investment would be below minimum levels even if a technical breakthrough could be made. Personnel from the Project Management Office (PMO) who were actively involved in the portfolio selection of projects also stated that they would never recommend approval of a project where the end result would have a shelf life of one year or less. Simply stated, these projects existed merely for the satisfaction of the two executives and to get them prestige with their colleagues.

Nevertheless, both executives found money for their projects and were willing to let them go forward without the standard approval process. Each executive was able to get an experienced project manager from his group to manage the pet project.

Gate-Review Meetings

At the first gate-review meeting, both project managers stood up and recommended that their projects be canceled and the resources assigned to other, more promising projects. They both stated that the technical breakthrough needed could not be made in a timely manner. Under normal conditions, both of these project managers should have received medals for their bravery in recommending that their projects be canceled. These recommendations certainly appeared to be in the best interests of the company.

But both executives were not willing to give up that easily. Canceling both projects would be humiliating for the executives who were sponsoring these projects. Instead, both executives stated that the projects were to continue until the next gate-review meeting, at which time a decision would be made for possible cancellation of both projects.

At the second gate-review meeting, both project managers once again recommended that their projects be canceled. And, as before, both executives asserted that the projects should continue to the next gate-review meeting before a decision would be made.

As luck would have it, the necessary technical breakthrough was finally made, but six months late. That meant that the window of opportunity to sell the products and recover the R&D costs would be six months rather than one year. Unfortunately, the thinking in the marketplace was that these products would be obsolete in six months, and no sales occurred for either product.

(Continued)

Case Study (Continued)

Both executives had to find a way to save face and avoid the humiliation of having to admit that they squandered a few million dollars on two useless R&D projects.

This could very well impact their year-end bonuses.

Questions

1. Is it customary for companies to allow executives to have pet or secret projects that do not follow the normal project approval process?

2. Who got promoted, and who got fired? In other words, how did the executives save face?

Reflections: The case outcomes could have been very different had the project managers been able to create experiences that show evidence of how the achievement of the outcomes of the two projects was not possible to reach. These experiences, coupled with the objective data sets, could have altered the course of these projects and expedited the achievement or an early cancellation, saving the organization large amounts of funds.

Tip

In creating the right experiencing, sponsors should act as role models for what the culture of an organization should use as a guide for how portfolio decisions are made.

2.3 Innovation Labs

The concept of innovation lab is an ideal way by which one could test the potential outcomes of experimenting without having to invest a large amount of time and resources. The innovation lab could be used for testing as many ingredients as necessary. This could be testing the way of working, the mix of the team, the go-to-market strategy, the right mix of partners, or the integration of multiple technologies.

In the ***irresponsible sponsor's*** case mentioned above, this concept could have been the missing ingredient for reaching a decision without having the executives look bad. The sponsors could have taken the lead on using the concept of an innovation lab to test the potential of achieving the technical breakthrough timely and thus could have looked like heroes and taken the credit their egos needed.

The other advantage of innovation labs is the streamlining of changes. As many organizations experience change fatigue, labs allow us to almost fully experience what is likely without having to go forward with the full scope of implementation. In the era of AI and digital twins, this becomes even more of a simpler task to execute.

In an interview with **Elham Nikookhesal, IKEA's Head of Project and Portfolio Management**, she answers the question: Why do you think that employees have the change fatigue you have noticed in IKEA?

Is it about trust, it is also about the articulation of the reasons behind the many changes that sometimes is not done well.

Having that long range map, NorthStar, helps in seeing where we are basically heading and then the key is to get the team convinced that this next change is okay to commit to. With the many changes, the team members are not motivated to follow the change message as they question becomes how the change champion would say if there weren't other sets of changes around the corner.

We are asking the team to be staying in this uncertain situation for an even longer time than we have already been positioning in previous change rounds.

Elham continued to add some description of the kind of culture that creates this success in better handling of changes.

Culture is important in driving our success. For the IKEA organization, this is the case, and it is important to the whole country, and it tied to our values, and so when you are so deep into creating the work results, you know the team values what you do. A key attribute to the success of our culture is driving consensus before you take key actions. You don't come and say, this is the right way of doing a certain thing and just move on. The key is to repeat the change message as many times as you can.

2.4 Creating a Connected Ecosystem

The need for a connected ecosystem has been increasingly relevant for how organizations work on creating a culture of openness and higher efficiencies that capitalizes on the multiple sensible and valuable areas of expertise and technological advances across stakeholders.

Looking across the ecosystem of an organization, like the one highlighted in the following case study, Zane Corporation, it is critical to have an understanding of the organization's culture, its dominant leadership behaviors, and the nature of the projects. The case will highlight the distinct differences between what is needed in operational projects versus innovation-type projects, which are critical examples of the experience-driven culture highlighted by this work.

Having a holistic strategic understanding of the business, the strategic objectives, and the demands around transformational and change initiatives is a fundamental quality for leading in these future cultures. In many cases, adapting and remaining flexible to utilize adjustable frameworks for executing work becomes the best scenario to choose. Having the innovation labs, discussed above, would allow project and program teams to test the applicability of some of the frameworks' choices before deciding on what to take on as a recommended approach.

Case Study

Zane Corporation[2]

Background

Zane Corporation was a medium-sized company with multiple product lines. More than 20 years ago, Zane implemented project management to be used in all their product lines, but mainly for operational or traditional projects rather than strategic or innovation projects. Recognizing that a methodology would be needed, Zane made the faulty conclusion that a single methodology would be needed and that a one-size-fits-all mentality would satisfy almost all their projects. Senior management believed that this would standardize status reporting and make it easy for senior management to recognize the true performance. This approach worked well in many other companies that Zane knew about, but it was applied to primarily traditional or operational projects.

As the one-size-fits-all approach became common practice, Zane began capturing lessons learned and best practices with the intent of improving the singular methodology. Project management was still being viewed as an approach for projects that were reasonably well defined, having risks that could be easily managed, and executed by a rather rigid methodology that had limited flexibility. Executives believed that project management standardization was a necessity for effective corporate governance.

The Project Management Landscape Changes

Zane recognized the benefits of using project management from their own successes, the capturing of lessons learned and best practices, and published research data. Furthermore, Zane was now convinced that almost all activities

2 Kerzner, 2022/John Wiley & Sons.

within the firm could be regarded as projects, and they were therefore managing their business by projects.

As the one-size-fits-all methodology began to be applied to nontraditional or strategic projects, the weaknesses in the singular methodology became apparent. Strategic projects, especially those that involved innovation, were not always completely definable at project initiation; the scope of work could change frequently during project execution; governance now appeared in the form of committee governance with significantly more involvement by the customer or business owner; and a different form of project leadership was required on some projects. Recognizing the true status of some of the nontraditional projects was becoming difficult.

The traditional risk management approach used on operational projects appeared to be insufficient for strategic projects. As an example, strategic projects require a risk management approach that emphasizes VUCA analyses:

- Volatility
- Uncertainty
- Complexity
- Ambiguity

Significantly more risks were appearing on strategic projects where the requirements could change rapidly to satisfy turbulent business needs. This became quite apparent on IT projects that focused heavily on the traditional waterfall methodology that offered little flexibility. The introduction of an agile methodology solved some of the IT problems but created others. Agile was a flexible methodology or framework that focused heavily on better risk management activities but required a great deal of collaboration. Every methodology or framework comes with advantages and disadvantages.

The introduction of an agile methodology gave Zane a choice between a rigid one-size-fits-all approach and a very flexible agile framework. Unfortunately, not all projects were perfect fits for an extremely rigid or flexible approach. Some were middle-of-the-road projects that fell in between rigid waterfall approaches and flexible agile frameworks.

Understanding Methodologies

Zane's original belief was that a methodology functioned as a set of principles that a company could tailor and then apply to a specific situation or group of activities that have some degree of commonality. In a project environment,

(Continued)

Case Study (Continued)

these principles might appear as a list of things to do and show up as forms, guidelines, templates, and checklists. The principles may be structured to correspond to specific project life-cycle phases.

For most companies, including Zane, the project management methodology, often referred to as the waterfall approach where everything is done sequentially, became the primary tool for the "command and control" of projects, providing some degree of standardization in the execution of the work and control over the decision-making process. Standardization and control came at a price and provided some degree of limitation as to when the methodology could be used effectively. Typical limitations that Zane discovered included:

Type of Project: Most methodologies assumed that the requirements of the project were reasonably well defined at the onset of the project. Trade-offs were primarily based on time and cost rather than scope. This limited the use of the methodology to traditional or operational projects that were reasonably well understood at the project approval stage and had a limited number of unknowns. Strategic projects, such as those involving innovation that had to be aligned to strategic business objectives rather than a clear statement of work, could not be easily managed using the waterfall methodology because of the large number of unknowns and the fact that they could change frequently.

Performance Tracking: With reasonable knowledge about the project's requirements, performance tracking was accomplished mainly using the triple constraints of time, cost, and scope. Nontraditional or strategic projects had significantly more constraints that required monitoring and therefore used other tracking systems than the project management methodology. Simply stated, the traditional methodology had limited flexibility when applied to projects that were not operational.

Risk Management: Risk management was important for all types of projects. However, on nontraditional or strategic projects, with the high number of unknowns that can change frequently over the life of the project, standard risk management practices that are included in traditional methodologies may be insufficient for risk assessment and mitigation practices.

Governance: For traditional projects, governance was provided by a single person acting as the sponsor for the project. The methodology became the sponsor's primary vehicle for command and control and was used with the mistaken belief that all decisions could be made by monitoring just the time, cost, and scope constraints.

Selecting the Right Framework

Zane recognized that the future was not simply a decision between waterfall, agile, and Scrum as to which one would be a best fit for a given project. New frameworks, perhaps a hybrid methodology, needed to be created from the best features of each approach and then applied to a project. Zane now believed with a reasonable degree of confidence that new frameworks, with a great deal of flexibility and the ability to be customized, would certainly appear in the future and would be a necessity for continued growth. Deciding which framework is best suited to a given project will be the challenge and project teams will be given the choice of which one to use.

Zane believed that project teams of the future would begin each project by determining which approach would best suit their needs. This would be accomplished with checklists and questions that address characteristics of the project, such as flexibility of the requirements, flexibility in the constraints, type of leadership needed, team skill levels needed, and the culture of the organization. The answers to the questions would then be pieced together to form a framework that may be unique to a given project.

Questions

1. What are some of the questions that Zane should ask themselves when selecting a flexible methodology?
2. What issues could arise that would need resolution?
3. What would you recommend as the first issue that needs to be addressed?
4. Was it a mistake or a correct decision not to allow the sales force to manage the innovation projects?
5. Is it feasible to set up a project management methodology for managing innovation projects?

Reflections: It is clear from the case study that maturing the portfolio and project management practices is a strategic priority for organizations. Building the openness in organizational cultures to have open dialogue about the differences in the types of projects, customers, and other stakeholders' expectations, and being open to adapting the choices that what fits, is critical. Organizations should build stronger sensing mechanisms to allow them to find out the unique attributes of transformation programs, especially those related to innovation, in order to make the proper choices that work across the ecosystem.

Tip
In creating the right future project execution framework, the culture should allow the initiatives' leaders the flexibility to choose the most fitting way of working for their teams.

2.5 Thinking Culture

One of the most critical dimensions in the experience-driven cultures of the future is their ability to become thinking cultures. It is almost one of the biggest challenges in leading organizations and projects today. The push for speed that we have been encountering and the preference of action over else have been such dominating behaviors across organizations. It is even seen as a critique point when leaders are seen as thinkers!

In the increasing reliance on digitalization and the ability to simulate just about anything in the virtual world, there is a hope that this will create a human 5.0 that is truly capable of thinking again for a change, just like the leadership guru John Maxell has been preaching for decades. If technology does not create the space, room, and time to think, then what true value we are gaining from it?

Thinking cultures are also the best in being learning cultures. They place premium on growth, challenging the norm, and put premium on creativity. All of these are critical to the type of innovation that is expected in the future. By dedicating time and building the muscle for thinking, leaders are able to do a better job reflecting on valuable lessons learned and conducting meaningful retrospectives in their initiatives. It has been said that Silicon Valley organizations use a "five-minute favor" method. Conceptually, this means that as anyone contributes ideas, there is a recommended five-minute window when only positive comments could be used. These types of approaches lead to scaling the impact of change and replicating success stories faster across boundaries in the future organization.

As shown in Figure 2.3, this thinking muscle will be highly complemented by data and AI capabilities, thus allowing humans to focus on the critical emoting required to connect across stakeholders and effectively drive change. The powerful use of computational unlimited potential, coupled with the care of tomorrow's leaders about our world, has the likelihood of tackling tomorrow's most complex missions.

In interviewing **Dr. Ed Hoffman, PMI Strategic Advisor** around culture topics, his points resonate well with the notion of the importance of thinking and its linkages to growth.

Figure 2.3 The Thinking Muscle. Credit: Enio-ia/Pixabay.

He shared the following related points:

> I think that driving future success, is enabled by cultures of growth, which are basically environments where curiosities speaking the truth is seen as a good thing, and where there is a commitment and resources that get behind the change. We got to have openness to learning at different levels and build on the freshness of new people coming on board and having innovative ideas.

To create experiences, I think cultures of growth basically support the trust required to experience lessons learned and to share openly. It leads to ongoing exchanges between young people and experienced people sitting around campfire talking, sharing, and creating the ability to learn and build.

The idea is that we talk about customers experience, put those ingredients in the culture, and really look at everything this created in terms of growth and learning. The activation of new ideas would create some of these experience-driven cultures in the future that we aspire to build. I think the organizations and societies in the countries that are successful, do have these values. With the fast changes, you must have a culture that's able to adapt more and value all the ideas that people have.

Tip

Thinking cultures are cultures of growth. Leaders place high value on building a growth mindset and making learning a critical organizational ethos ingredient.

Reference

Kerzner, H. and Zeitoun, A. (2023). Cracking the Excellence Code.

Review Questions

Parentheses ( ) are used for Multiple Choice when one answer is correct. Brackets [ ] are used for Multiple Answers when many answers are correct.

1 Which of the following is preferred in selecting a way of working that allows leaders to address the work ahead of the team?
( ) An approach that works across all teams and complexity levels.
( ) A fitting way of working.
( ) Good one-size-fits-all process.
( ) A framework that uses all AI capabilities.

2 What are good ingredients for future excellence building blocks? Choose all that apply.
[ ] Practicing quality questions.
[ ] Extensive planning.
[ ] Building the delivery muscle.
[ ] Questioning coupled with experimenting.

3 What is the greatest value of having a responsible sponsor?
( ) Ensures performance reporting is a top priority.
( ) Taking risks across all situations.
( ) Making strategic choices that are built on a clear business case.
( ) Makes choices that suit their interests.

4 What is the innovation lab concept?
( ) It only applies to testing technologies before they are released.
( ) It is an opportunity to slow things down.
( ) Making choices based on experimenting opportunities.
( ) A mechanism for applying research to good use.

5 Reviewing the Zane Corporation case study, what is a key learning in deciding on a project management framework? Choose all that apply.
[ ] Practicing adaptability in decision-making.
[ ] Only listen to the project teams.
[ ] Building a framework that can adjust with relevant factors.
[ ] Question everyone's motives.

6 What is the most concern with innovation initiatives in relation to operational projects?

( ) They are longer duration.

( ) The degree of risk and uncertainty leading to dynamic scope expectations.

( ) They tend to be less strategic.

( ) They require higher amount of performance reporting.

7 What is a possible by-product of building a thinking culture?

( ) They take so much time to make decisions.

( ) Slowing down to go faster.

( ) They tend to be tactically focused.

( ) They require no change in what gets rewarded.

3

Driving Integration

Future leaders are integrators of people, processes, systems, and technology. These are the key ingredients for transformation, and the future continues to be built on transformative ideas that allow scaling, address most complex challenges, and enable us to achieve sustainability ambitions.

Driving integration across experience-driven cultures requires working collaboratively across the ecosystem. It assumes a level of openness across the enterprise, the customers, partners, suppliers, and an extended body of stakeholders. Driving integration is an organizational muscle that should cascade across levels of the organization.

Having the ability to see how the pieces connect, how the work elements contribute to a change roadmap, and how the various roadmaps connect to achieving the vision of the organization are all examples of how integration should prevail. These connections enable the organization to act and behave as one body in front of its customers and to create rich experiences for employees, management, stakeholders, shareholders, and others affected by the outcomes of the enterprise.

Key Learnings

- Understand, using a case example, the importance of establishing clear portfolio, program, and project links.
- Get immersed in the portfolio management professional certification and its value to the enterprise.
- Learn key views on how to be more disciplined around the achievement of value.
- Learn the art of simplicity in connecting complex topics and working across groups of stakeholders, using the Program Way.
- Develop your integrating storytelling capabilities while using data insights to empower the story's impact.

Creating Experience-Driven Organizational Culture: How to Drive Transformative Change with Project and Portfolio Management, First Edition. Al Zeitoun.

3.1 The Portfolio, Program, and Project Link

For the leaders to become such effective integrators, there is a natural way to do this in the art of connecting portfolios, programs, and projects. This natural connection allows you as leaders to understand the importance of integration by looking end-to-end, developing better system integration muscles, and creating the objective views necessary to experience effective decision-making. Figure 3.1 is a simple illustration that shows how portfolios could represent the integration of other portfolios, various programs, and related projects.

The value proposition of managing business across organizational cultures using this portfolio management format is to have clear visibility of where our investments are spread, where resources are allocated, and how to best capitalize on the energy and collaboration of the right key stakeholders. Having clear connections across the elements of the portfolio goes a long way in achieving efficiencies that are otherwise left uncapitalized when managed disjointly.

> **Tip**
> In building connected organizational cultures, it is valuable to use portfolio management structures that clearly connect with the associated programs and projects.

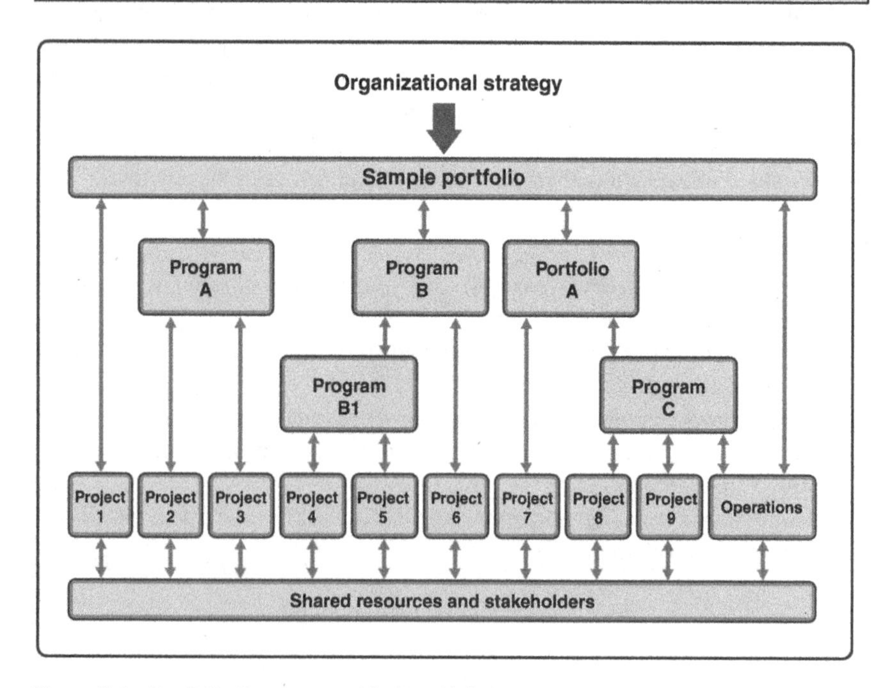

Figure 3.1 Portfolio, Program, and Project Linkages.

The following case study highlights elements of this crucial link. An acquisition initiative is potentially a complex portfolio of multiple programs and projects. If not approached correctly with a clear lifecycle view of managing that portfolio, it is likely that critical dimensions will be forgotten, especially given the pace by which the organizations get into it, or the lack of realization of potential key cultural differences, or the resistance that could exist between the organizations that are part of that initiative.

The discipline of project and program management is highly critical to the success of many of these strategic initiatives. Having the proper and clear portfolio links established is also a valuable mechanism in handling potential risks and avoiding getting into critical issues that dilute the potential of the long-term value.

Case Study

Acquisition Problem[1]

Background

All companies strive for growth. Strategic plans are prepared by identifying new products and services to be developed and new markets to be penetrated. Many of these plans require mergers and acquisitions to obtain the strategic goals and objectives rapidly. Yet often, even the best-prepared strategic plans fail when based on mergers and acquisitions. Too many executives view strategic planning for a merger or acquisition as planning only and often give little consideration to implementation, which takes place when both companies are actually combined. Implementation success is vital during any merger and acquisition process.

Planning for Growth

Companies can grow in two ways: internally or externally. With internal growth, companies cultivate their resources from within and may spend years attaining their strategic targets and marketplace positioning. Since time may be an unavailable luxury, meticulous care must be given to make sure that all new developments fit the corporate project management methodology and culture.

External growth is significantly more complex. It can be obtained through mergers, acquisitions, and joint ventures. Companies can purchase the expertise they need very quickly through mergers and acquisitions. Some companies

(Continued)

1 Kerzner, 2022/John Wiley & Sons.

Case Study (Continued)

execute occasional acquisitions, while other companies have sufficient access to capital such that they can perform continuous acquisitions. However, once again, companies often neglect to consider the impact on project management after the acquisition is made. Best practices in project management may not be transferable from one company to another. The impact on project management systems resulting from mergers and acquisitions is often irreversible, whereas joint ventures can be terminated.

Project management often suffers after the actual merger or acquisition. Mergers and acquisitions allow companies to achieve strategic targets at a speed not easily achievable through internal growth, provided the sharing or combining of assets and capabilities can be done quickly and effectively. This synergistic effect can produce opportunities that a firm might be hard-pressed to develop by itself.

Mergers and acquisitions focus on two components: pre-acquisition decision-making and post-acquisition integration of processes. Wall Street and financial institutions appear to be more interested in the near-term financial impact of the acquisition rather than the long-term value that can be achieved through combined or better project management and integrated processes. During the mid-1990s, companies rushed into acquisitions in less time than the company required for a capital expenditure approval. Virtually no consideration was given to the impact on project management or whether project management knowledge and best practices would be transferable.

The result appears to have been more failures than successes. When a firm rushes into an acquisition, often very little time and effort are spent on post-acquisition integration. Yet this is where the real impact of the acquisition is felt. Immediately after an acquisition, each firm markets and sells products to each other's customers. This may appease the stockholders, but only in the short term. In the long term, new products and services will need to be developed to satisfy both markets. Without an integrated project management system where both parties can share the same intellectual property and work together, this may be difficult to achieve.

When sufficient time is spent on pre-acquisition decision-making, both firms look at combining processes, sharing resources, transferring intellectual property, and the overall management of combined operations. If these issues are not addressed in the pre-acquisition phase, then the unrealistic expectations may lead to unwanted results during the post-acquisition integration phase.

Strategic Timing Issue

Lenore Industries had been in existence for more than 50 years and served as a strategic supplier of parts to the automobile industry. Lenore's market share was second only to its largest competitor, Belle Manufacturing. Lenore believed that the economic woes of the U.S. automobile industry between 2008 and 2010 would reverse themselves by the middle of the next decade and that strategic opportunities for growth were at hand.

The stock prices of almost all of the automotive suppliers were grossly depressed. Lenore's stock price was also near a 10-year low. But Lenore had rather large cash reserves and believed that the timing was right to make one or more strategic acquisitions before the marketplace turned around. With this in mind, Lenore decided to purchase its largest competitor, Belle Manufacturing.

Pre-acquisition Decision-Making

Senior management at Lenore fully understood that the reason for most acquisitions is to satisfy strategic and/or financial objectives. Table 3.1 shows the six reasons identified by senior management at Lenore for the acquisition of Belle Manufacturing and the most likely impact on Lenore's strategic and financial objectives. The strategic objectives are somewhat longer-term than the financial objectives, which are under pressure from stockholders and creditors for quick returns.

Table 3.1 Acquisition Objectives.

Reason for Acquisitions	Strategic Objective	Financial Objective
Increase customer base	Bigger market share	Bigger cash flow
Increase capabilities	Become a business solution provider	Larger profit margins
Increase competitiveness	Eliminate costly steps and redundancy	Stable earnings
Decrease time to market for new products	Market leadership	Rapid earnings growth
Decrease time to market for enhancements	Broad product lines	Stable earnings
Closer to customers	Better price–quality–service mix	Sole-source or single-source procurement

(Continued)

Case Study (Continued)

Lenore's senior management fully understood the long-term benefits of the acquisition, which were:

- Economies of combined operations.
- Assured supply or demand for products and services.
- Additional intellectual property, which may have been impossible to obtain otherwise.
- Direct control over cost, quality, and schedule rather than being at the mercy of a supplier or distributor.
- Creation of new products and services.
- Putting pressure on competitors by creating synergies.
- Cutting costs by eliminating duplicated steps.

Lenore submitted an offer to purchase Belle Manufacturing. After several rounds of negotiations, Belle's board of directors and Belle's stockholders agreed to the acquisition. Three months later, the acquisition was completed.

Post-acquisition Integration

The essential purpose of any merger or acquisition is to create lasting value and value that would not exist had the companies remained separate. The achievement of these benefits, as well as attaining the strategic and financial objectives, could rest on how well the project management value-added chains of both firms are integrated, especially the methodologies within their chains. Unless the methodologies and cultures of both firms can be integrated, and reasonably fast, the objectives may not be achieved as planned.

Lenore's decision to purchase Belle Manufacturing never considered the compatibility of their respective project management approaches. Project management integration failures occurred soon after the acquisition happened. Lenore had established an integration team and asked the integration team for a briefing on what critical issues were preventing successful integration.

The integration team identified five serious problems that were preventing successful integration of their project management approaches:

1) Lenore and Belle have different project management methodologies.
2) Lenore and Belle have different cultures, and integration is complex.
3) There are wage and salary disparities.
4) Lenore overestimated the project management capability of Belle's personnel.
5) There are significant differences in functional and project management leadership.

It was now apparent to Lenore that these common failures resulted because the acquisition simply cannot occur without organizational and cultural changes that are often disruptive in nature. Lenore had rushed into the acquisition with lightning speed but with little regard for how the project management value-added chains would be combined.

The first common problem area was inability to combine project management methodologies within the project management value-added chains. This occurred for the following four reasons:

1) A poor understanding of each other's project management practices prior to the acquisition.
2) No clear direction during the pre-acquisition phase on how the integration would take place.
3) Unproven project management leadership in one or both firms.
4) The existence of a persistent attitude of "we–them."

Some methodologies may be so complex that a great amount of time is needed for integration to occur, especially if each organization has a different set of clients and different types of projects. As an example, a company developed a project management methodology to provide products and services for large publicly held companies. The company then acquired a small firm that sold exclusively to government agencies.

The company realized too late that integration of the methodologies would be almost impossible because of requirements imposed by government agencies for doing business with the government. The methodologies were never integrated, and the firm servicing government clients was allowed to function as a subsidiary with its own specialized products and services. The expected synergy never took place.

Some methodologies simply cannot be integrated. It may be more prudent to allow the organizations to function separately than to miss windows of opportunity in the marketplace. In such cases, pockets of project management may exist as separate entities throughout a large corporation.

Lenore knew that Belle Manufacturing services many clients outside of the United States but did not realize that Belle maintained a different methodology for those clients. Lenore was hoping to establish just one methodology to service all clients.

The second major problem area was the existence of differing cultures. Although project management can be viewed as a series of related processes, it is the working culture of the organization that must eventually execute these processes. Resistance by the corporate culture to effectively support

(Continued)

Case Study (Continued)

project management can cause the best plans to fail. Sources for the problems with differing cultures include a culture that:

- Has limited project management expertise (i.e., missing competencies) in one or both firms.
- Is resistant to change.
- Is resistant to technology transfer.
- Is resistant to transfer of any type of intellectual property.
- Will not allow for a reduction in cycle time.
- Will not allow for the elimination of costly steps.
- Must reinvent the wheel.
- Views project criticism as personal criticism.

Integrating two cultures can be equally difficult during favorable and unfavorable economic times. People may resist any changes to their work habits or comfort zones, even though they recognize that the company will benefit from the changes.

Multinational mergers and acquisitions are equally difficult to integrate because of cultural differences. Several years ago, an American automotive supplier acquired a European firm. The American company supported project management vigorously and encouraged its employees to become certified in project management. The European firm provided very little support for project management and discouraged its workers from becoming certified, arguing that its European clients do not regard project management as highly as do General Motors, Ford, and Chrysler.

The European subsidiary saw no need for project management. Unable to combine the methodologies, the American parent company slowly replaced the European executives with American executives to drive home the need for a single project management approach across all divisions. It took almost five years for the complete transformation to take place. The American parent company believed that the resistance in the European division was more of a fear of change in its comfort zone than a lack of interest by its European customers.

Planning for cultural integration can also produce favorable results. Most banks grow through mergers and acquisitions. The general practice in the banking industry is to grow or be acquired. One Midwest bank recognized this and developed project management systems that allowed it to acquire other banks and integrate the acquired banks into its culture in less time than other banks allowed for mergers and acquisitions. The company viewed project management as an asset that had a very positive effect on the corporate bottom line. Many banks today have manuals for managing merger and acquisition projects.

The third problem area Lenore discovered was the impact on the wage and salary administration program. The common causes of the problems with wage and salary administration included:

- Fear of downsizing
- Disparity in salaries
- Disparity in responsibilities
- Disparity in career path opportunities
- Differing policies and procedures
- Differing evaluation mechanisms

When a company is acquired and integration of methodologies is necessary, the impact on wage and salary administration can be profound. When an acquisition takes place, people want to know how they will be affected individually, even though they know that the acquisition is in the best interests of the company.

The company being acquired often has the greatest apprehension about being lured into a false sense of security. Acquired organizations can become resentful to the point of trying to subvert the acquirer. This will result in value destruction where self-preservation becomes paramount, often at the expense of project management systems.

Consider the following situation. Company A decides to acquire Company B. Company A has a relatively poor project management system, where project management is a part-time activity and not regarded as a profession. Company B, in contrast, promotes project management certification and recognizes the project manager as a full-time, dedicated position. The salary structure for the project managers in Company B was significantly higher than for their counterparts in Company A. The workers in Company B expressed concern that "We don't want to be like them," and self-preservation led to value destruction.

Because of the wage and salary problems, Company A tried to treat Company B as a separate subsidiary. But when the differences became apparent, project managers in Company A tried to migrate to Company B for better recognition and higher pay. Eventually, the pay scale for project managers in Company B became the norm for the integrated organization.

When people are concerned with self-preservation, the short-term impact on the combined value-added project management chain can be severe. Project management employees must have at least the same, if not better, opportunities after acquisition integration as they did prior to the acquisition.

(Continued)

Case Study (Continued)

The problem area that the integration team discovered was the overestimation of capabilities after acquisition integration. Included in this category were:

- Missing technical competencies
- Inability to innovate
- Speed of innovation
- Lack of synergy
- Existence of excessive capability
- Inability to integrate best practices

Project managers and those individuals actively involved in the project management value-added chain rarely participate in pre-acquisition decision-making. As a result, decisions are made by managers who may be far removed from the project management value-added chain and whose estimates of post-acquisition synergy are overly optimistic.

The president of a relatively large company held a news conference announcing that their company was about to acquire another firm. To appease the financial analysts attending the news conference, they meticulously identified the synergies expected from the combined operations and provided a timeline for new products to appear on the marketplace. This announcement did not sit well with the workforce, who knew that the capabilities were overestimated and the dates were unrealistic. When the product launch dates were missed, the stock price plunged, and blame was erroneously placed on the failure of the integrated project management value-added chain.

In this case, the problem area identified was leadership failure during post-acquisition integration. Included in this category were:

- Leadership failure in managing change
- Leadership failure in combining methodologies
- Leadership failure in project sponsorship
- Overall leadership failure
- Invisible leadership
- Micromanagement leadership
- Believing that mergers and acquisitions must be accompanied by major restructuring

Managed change works significantly better than unmanaged change. Managed change requires strong leadership, especially with personnel experienced in managing change during acquisitions.

Company A acquires Company B. Company B has a reasonably good project management system, but it has significant differences from Company A's system. Company A then decides, "We should manage them like us," and nothing should change. Company A then replaces several Company B managers with experienced Company A managers, a change that took place with little regard for the project management value-added chain in Company B. Employees within the chain in Company B were receiving calls from different people, most of whom were unknown to them and were not told whom to contact when problems arose.

As the leadership problem grew, Company A kept transferring managers back and forth. This resulted in smothering the project management value-added chain with bureaucracy. As expected, performance was diminished rather than enhanced, and the strategic objectives were never attained.

Transferring managers back and forth to enhance vertical interactions is an acceptable practice after an acquisition. However, it should be restricted to the vertical chain of command. In the project management value-added chain, the main communication flow is lateral, not vertical. Adding layers of bureaucracy and replacing experienced chain managers with personnel inexperienced in lateral communications can create severe roadblocks in the performance of the chain.

The integration team then concluded that any of the problem areas, either individually or in combination, could cause the project management value chain to have problem areas, such as:

- Poor deliverables
- Inability to maintain schedules
- Lack of faith in the chain
- Poor morale
- Trial by fire for all new personnel
- High employee turnover
- No transfer of project management intellectual property

Company A now realized that it may have bitten off more than it could chew. The problem was how to correct these issues in the shortest amount of time without sacrificing its objectives for the acquisition.

(Continued)

Case Study (Continued)

Questions

1. Why is it so difficult to get senior management to consider the impact on project management during pre-acquisition decision-making?
2. Are the acquisition objectives in Table 3.1 realistic?
3. How much time is really needed to get economies of combined operations?
4. How should Lenore handle differences in the project management approach if Lenore has the better approach?
5. How should Lenore handle differences in the project management approach if Belle has the better approach?
6. How should Lenore handle differences in the project management approach if neither Lenore nor Belle has any project management?
7. How should Lenore handle differences in the culture if Lenore has a better culture?
8. How should Lenore handle differences in the culture if Belle has the better culture?
9. How should Lenore handle differences in the wage and salary administration program?
10. Is it possible to prevent an overoptimistic view of the project management capability of the company being acquired?
11. How should Lenore handle disparities in leadership styles?

Tip

Approaching key strategic or transformation initiatives requires a portfolio discipline that balances culture, leadership, and the fitting project management framework.

3.2 The Portfolio Management Professional

Certification is one of these topics that seems to have ongoing open debates across organizations and cultures. I still remember the time when I became a Project Management Professional as I lived in Wichita, Kansas, in the U.S. at the time and was the first certified in that state.

Although having been a practitioner for some time, the value that certification added is the commitment to the disciple and to the profession. It is also a brand creator that got me positioned for higher responsibilities and the courage to cross industries and apply similar practices within these new turfs.

The debate is typically around questioning true value of certification and whether the certified practitioner is truly more effective than others who are not. Of course, there is no assurance that this will be the case. It is ultimately the individual and their commitment to learning, experiencing, and growing the toolbox of capabilities and skills and the associated behaviors.

The Portfolio Management Professional (PfMP) is no exception. In fact, it is the type of certification that confirms that the practitioner is in the position of being able to take more strategic responsibilities, as it is the closest to the level of executing against the strategic aspirations of an organization.

This is the certification that is focused on testing the ability of the practitioner of being able to make strategic choices across the portfolio management lifecycle. It is about ensuring that we do the right work and less about the mechanics of getting the work done right, as in the case of the project manager's role.

As seen in the abovementioned acquisition case study, the capabilities that a PfMP could have brought to the surface could dramatically increase the chances of looking at such acquisitions more strategically, balancing short- and long-term benefits. It is expected that this professional is able to invest the right rigor upfront to make choices that are meaningful to the organization. This professional looks at the cultural implications of such acquisitions and is able to understand how to implement the right strategies of working across stakeholders and influencing key players to make the right decisions and take the right actions timely.

The combination of years of portfolio management experience, the review and evaluation of a panel, and the certification exam give the certification the necessary rigor for this professional to be proud of such an accomplishment and offer the organizations the confidence that this leader is capable of operating strategically to handle diverse and complex portfolio of strategic, transformational, and operational initiatives.

As seen in Figure 3.2, and as in the case of ISO Standards, the various Project Management Institute (PMI) standards and approaches to certification follow similar rigor through the American National Standards Institute (ANSI).

Figure 3.2 The Certified Professional. Credit: u_4xcm1iw8y9/Pixabay.

> **Tip**
> Approaching certification with a thirst for knowledge and keen interest in enhancing one's practices is a good beginning. The changes in behaviors and outcomes that follow are key.

3.3 The Value Focus

Following the Program Way highlighted above, and as the maturity of organizations increases, the chances grow for establishing the right commitment to value achievement. Doing business with integration at the core supports building this value focus. This is such a critical distinction in cultures that are experience-driven. In these cultures, there is clarity that we can't be tactical only about what success or value looks like. It can't just be one thing or the other. It has to be holistic. For example, the success of an initiative can't just be measured by timely achievement of scope within a certain budget.

Many initiatives could focus on many of these tactical metrics and miss the boat on achieving what true value is and what truly matters to customers and other key stakeholders. This mindset shift is critical. Driving work with that integrating view is key. This manifests itself in the mix of metrics that one chooses. The mix

Figure 3.3 The Value Focus. Credit: u_4xcm1iw8y9/Pixabay.

should not include just classical metrics, but also value ones, and few intangibles, thus enabling us to have a set of strategic metrics that matter.

One of the most critical aspects of creating a value focus is not only focus on the short term as seen, for example, by increased profits as highlighted by Figure 3.3 or as reflected in the acquisition case study mentioned above that could be influenced by pressure of the market or shareholders to financial value quickly. The leader should have the ability to be courageous and push back on the myopic view of value and be in a position to ask the right strategic questions timely of the proper stakeholders in order to achieve a unified and balanced view of success.

It is also critical in the value focus to have the fluidity to reassess value across the lifecycle. Cultures that experience skills, products, innovations, and people are more likely to understand this and remain adaptable to changing course and rethink what is necessary to achieve a new view of success and value. This also includes managing the element of perception that is typically associated with how value is seen by affected stakeholders.

3.4 Integrating with Simplicity

One of the most common ways to integrate is the program mindset. This could be referred to as the Program Way. In our article, Kerzner and Zeitoun (2023), we tackled the topic of programs and how they enable organizations and their program teams to integrate across various initiatives and simplify the achievement of value. Simplicity is an art and is especially critical in the case of programs that combine a vast number of stakeholders and cross multiple boundaries within the ecosystem.

3.4.1 Introduction

A program is generally defined as a grouping of projects that can be managed consecutively or concurrently or a combination of both. There are numerous challenges facing the program manager that quite often make it difficult to achieve all or even part of the strategic goals and objectives established by senior management. The larger and more complex the program, the more difficult it will be to overcome the challenges.

Many of the challenges are common to both projects and programs. However, the risks due to the challenges may have a much greater impact on programs than projects. When projects are challenged, some companies simply let the project fail, and the team moves on to their next project assignment. When programs are challenged, the cost of terminating a program can be quite large and might have a serious impact on the organization's competitiveness and future success.

Projects generally have a finite time duration. Most programs, because of their strategic nature and impact on the success of the organization, are much longer in duration and susceptible to more challenges, risks, and negative impacts on the business.

In the early years of project management, most PMs had engineering backgrounds, many with advanced degrees in technical disciplines. Project sponsors were assigned from the senior-most levels of management, mainly to make all of the necessary business-related and strategic decisions. Many companies did not trust project managers to make business or strategic decisions. Even in companies that had programs and program managers, there were still governance personnel assigned to ensure linkages to strategic business objectives.

Project management today is more than just a traditional career path for workers. It is now treated as a strategic competency, which means it is one of the 4 or 5 most important career paths in the company in order for the firm to have a viable and successful future. Part of the strategic competency requires that senior management give up the idea that information is power and clearly share strategic information with project managers. Today's project managers and program managers are managing strategic opportunities for companies and making strategic decisions. This forms the shift to the Program Way, running projects and programs with the proper strategic clarity and full authorization to make the necessary business strategic value decisions, a true business strategist way.

The program business case must articulate the expected benefits and business value. The business case also provides the boundaries for many of the decisions that will have to be made. The challenge will be in the preparation of a business case such that all program team members clearly understand what is expected of them.

Program management is more closely aligned to strategic decisions than project management activities that focus on traditional projects. As such, over the next decade, we can expect to see a significant growth in the "Program Way," with program managers becoming experts in strategic planning.

Several years ago, IBM wanted all of their project managers to become dual certified: certified by PMI on project management and certified internally by IBM in the use of IBM's forms, guidelines, templates, and checklists for making strategic decisions at IBM. IBM discovered the importance of having their 46,000 PMs qualified and trained in making business decisions on projects and programs.

Other companies have followed IBM's lead and created internal training and internal certification programs more closely aligned to business strategy. Even without utilizing the words, this expansion of capability building toward business strategy linkages most certainly confirms that companies have been shifting focus to expand and make use of the "Program Way."

3.4.2 The Program Way

Program stakeholders are the people who ultimately decide whether a program is successful. There can be significantly more stakeholders in programs than in projects. Failing to meet program stakeholder expectations can result in a significant loss of business. Given the long-time frame of many programs, managing the changes in stakeholders over the lifecycle of the program, and addressing their changing expectations, this ***Program Way*** muscle is critical to develop.

To support this ***Program Way***, dedicated program personnel would likely be required, such as:

- **Program Office Manager**: This can include handling administrative paperwork, meeting scheduling, and making sure that program activities are aligned to company standards and expectations.
- **Reports Manager**: This person is responsible for the preparation and distribution of all reports and handouts. The person is usually not involved in the actual writing of the reports. Naturally, there is an opportunity here to exploit the power of artificial intelligence for these first two roles.
- **Risk Manager**: This person monitors the Volatility, Uncertainty, Complexity, and Ambiguity (VUCA) environment and the enterprise environment factors. Additional responsibilities include risk identification, analysis, and response to all risks that can impact program success.

- **Business Analyst**: This person works closely with the risk manager, and activities may include identification of business opportunities and threats. The analyst may monitor compliance with customer requirements and verification of the program's deliverables.
- **Change Manager**: Some large programs may clearly indicate that changes in the firm's business model will be necessary. The change manager prepares the organization for the expected change. The change may just be in some of the processes or the way that the firm conducts its business rather than a significant change to the business model.

3.4.3 The Program Way Letter to Future Program Managers

There are both concerns and equally a level of excitement about the future of the Program Way with the amount of disruption and the anticipated changing environmental and business dynamics. As a program manager, you are in the right place at the center of leading through chaos and creating opportunities. This brief letter to future program managers is intended to highlight some of the key anticipated shifts ahead, be aware of them, prepare for them, and ultimately put that readiness to good use in creating meaningful strategic impact.

The changing nature of your role and possibly title. Whether program manager remains as a title or gets replaced with some elements of leading, collaborating, strategizing, integrating, coaching, or driving, it is all about creating impact. There is a dominant need for servant or social leadership, where the program leader is able to adapt between being the coach or becoming the one carrying the program team across obstacles.

These critical changes are shaping you to be the true organizational connector. Your role will continue to balance technology and strategy as your key enablers. Program managers have to have their voice in working across the business boundaries and continually breaking down actual and mental silos in the organization.

Future program managers are connectors.
Readiness for that future also has growth and people components. The open window for continual learning is a feature that is strengthened by technology and artificial intelligence. This requires developing an appetite for your growth and for equally growing other key stakeholders around you. A more mature and developed stakeholder community directly contributes to making your future role most effective.

In the future, program managers will also have a vital impact on sustaining the growth of businesses and people. With the norm shifting to program managers being more aligned with the executive teams, being part of the most critical strategic dialogues and decisions, and having the right seat at the table, the value of program management continues to become more evident. This makes your role even more clearly strategic in terms of impact-driving and effecting the future of organizations.

3.4.4 The Program Way's Future Conductor

Describing the role of the program manager could take many forms. One of the favorites could be the orchestra conductor. The anticipated Program Way's integration emphasis played in the role of the program manager, and the need to align across a diverse set of stakeholders, make the conductor analogy a strong fitting designation.

One of the experiences worth discovering would be to get the opportunity to go deep in understanding the role of the orchestra conductor and see that, although it is such a critical role, the quality of the outcome of that musical piece's delivery ultimately rests with the orchestra itself, its training, and its achieved harmony.

- The role of a conductor is to unify a large group of musicians into a core sound instead of a wild bunch of different sounds surging out.
- The program becomes similar to the nicely played piece like the one we enjoy going to the theater for.
- Sees what good looks like.
- Stepping back and seeing the cross-dynamics.
- Bringing the team toward benefits and strategic outcomes.

3.4.5 The Holistic View of the Program Way

An analogy that relates to the importance of the strategic and holistic view of the ***Program Way*** is being on the balcony versus being on the dance floor, as illustrated in Figure 3.4. Leaders realize the importance of this. This contributes to the adaptable mindset that program managers should possess.

Program managers tend to be more successful when they have the ability to create the distance and see better where some of the gaps might be in what is happening in front of their eyes on the dance floor or in the deep work the program team is involved with. They should also have the ability to roll up their sleeves and jump right back into it and be in the trenches with their program team colleagues.

3.4.6 The Path Forward

Shifts in how we work and how we deliver value require developing a sensible and holistic new program muscle. The ***Program Way*** is about excelling in building

Figure 3.4 The Holistic View. Credit: u_4xcm1iw8y9/Pixabay.

the links between initiatives and the strategic imperatives behind those initiatives. Executives need to understand and demonstrate the shift to this way of working and entrust their program managers to drive the achievement of outcomes from their most significant business initiatives.

The ***Program Way*** requires the creation of an intentionally strategic culture. Just like PMI and other global organizations have been directing the attention toward a focus on benefits and value, businesses should be ready to support the shift needed to grow the business in the buckets of programs that fully align with clear *Strategic Focus Areas*. This is our opportunity to raise the bar around the practices of project and program management and ensure that the right organizational champions provide the proper attention to growing the *Next Gen* of program leaders to take the helm of the most critical future transformational changes and the associated organizational leadership roles.

Tip

In integrating with simplicity, one should consider the ***Program Way***. Programs are best suited to connect the dots across the work elements of key initiatives.

3.5 Integrating Stories

Stories are powerful. Sometimes it is also, the less is more concept. As leaders in tomorrow's organizations, creating the right experience-driven culture hinges on great stories where people can connect and be inspired. Stories allow program and project teams to be connected to a purpose that matters. Many of today's challenges of achieving high-performing results on teams link to missing motivation or clarity. When there is the right story and clarity is achieved, people connect and the extended groups of stakeholders can relate. This is also where change could materialize and scale.

Let's go back to the movies and pick "**Tommy Boy**." In a turn of events, the dad of Tommy dies, leaving him with the responsibility of saving the manufacturing plant that the father has run successfully for years. Tommy has not been in such a leading spot and has tended to that point to be just focused on having a party and good time in the company of not-so-great friends. All the odds were against him to step into the shoes of his dad, save a factory, jobs, and the livelihood of a town that depended on this.

Tommy tried to step in, yet he lacked the understanding, expertise, and ability to connect to potential clients needed to keep the business running. Even though he was assigned a dedicated employee from the factory who has good expertise about the business, this did not seem to help. The brand of the factory seemed to have been linked to the father and his style and image, so having the same last name alone did not seem to help.

To reflect on the culture in that factory first before looking to see how Tommy manages to achieve a transformation, one would wonder if the dad did right by the organization. Creating an experience-driven culture would have meant establishing an environment where all voices are heard, joint commitment is thriving, and creative minds are nourished in preparation for such risky forks on the road.

It did not seem to have been a dedication to creating an environment where leadership practices were thriving or where other leaders could naturally step up and lead when needed. Tommy, even as a likely successor, was definitely not prepared for such a moment.

The movie continues to highlight the failing journey of Tommy attempting to get through to potential clients and finding extreme ways to get trust and credibility established. Only when something clicked inside him, maybe due to the continued sarcasm or lack of belief in him on the part of his road trip partner, that he manage to dig deep into his abilities and possible hidden talents. He was able to find the missing link. It was about articulating the right effective story.

Tommy was able to formulate stories that were simple, connect, and inspire other organizational leaders to commit to ordering their needs for brakes from

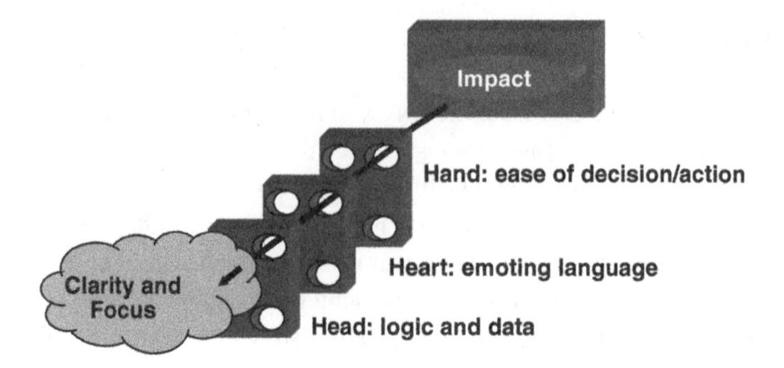

Figure 3.5 The Integrating Stories Journey.

the factory. He was able to use integration as a weapon. Integrating stories connect the logic with the emotion. Tommy integrated logic about the known quality of the products of the factory to the emotional needs of the customers in producing automobiles that are safe and clearly linked that safety to their end customers. He was able to use proper emoting language in his stories, which even connected the clients to the sense of safety that they wanted to secure for their own families and friends. Establishing as many of these bridges in communication is at the center of the art of developing and delivering integrating stories.

In an organization that is seeking to create the right experiences for its employees, customers, and extended stakeholders, there needs to be a commitment to ensuring that the brand, the offerings, and the stories used to describe the value of these offerings are all connected. When everyone is connected to the experience-driven environment, then all are in a position to best represent the organization and its solutions.

Just like in the case of Tommy Boy, integrating stories seem to have a number of common attributes that make them effective. These attributes usually span the 3Hs, the head, heart, and hand. Figure 3.5 highlights these attributes in the integrating journey of delivering a well-planned story from the beginnings of simplicity and clarity to the destination of creating the proper impact while crossing the stages of the head, heart, and hand.

Tip

In designing the powerful integrating stories, leaders should establish the right fluid connections between the head, heart, and hand.

Reference

Kerzner, H. and Zeitoun, A. (2023). The program way, the great project management accelerator, series article. *PM World Journal* XII: Issue VIII, August.

Review Questions

Parentheses () are used for Multiple Choice when one answer is correct. Brackets [] are used for Multiple Answers when many answers are correct.

1 Which of the following is an example of what a portfolio management view enables toward the success of strategic initiatives?
- ( ) Increased use of digital solutions.
- ( ) Seeing the interdependencies across portfolio lifecycle decisions.
- ( ) It helps us slow down the decision-making process.
- ( ) A framework that has to be applied across all portfolio sizes.

2 What are good ingredients for successful acquisition initiatives? Choose all that apply.
- [ ] Pre-work that covers proper portfolio initiation steps.
- [ ] Likely amount of executives' bonus.
- [ ] Balancing the short-term with the long-term benefits.
- [ ] Paying attention to the culture fit and how it affects post-acquisition work.

3 What is the greatest value of having the right unified project management framework for acquisitions?
- ( ) Ensures applying the right disciple for the steps that are needed post-acquisition.
- ( ) Applies enough policing to give management reports needed.
- ( ) Making project managers busy.
- ( ) Enabling the merged organizations to choose whatever they deem a good way for them.

4 What is the value of being a certified PfMP?
- ( ) It adds more post-nominal letters to one's name.
- ( ) It is an opportunity to highlight how certain leaders are more capable than others.

(Continued)

(Continued)

( ) Making a clear commitment to applying the principles toward strategic outcomes.
( ) Showing off one's capabilities.

5 What are ways to mature the value focus in cultures that seek to mature their experiencing practices? Choose all that apply.
[ ] Include intangibles in the set of chosen metrics.
[ ] Only listen to what the customer says.
[ ] Build a success measurement framework that targets having strategic metrics.
[ ] Focus on getting the fundamental metrics covered.

6 What is a key focus of the Program Way?
( ) Program managers are no longer needed.
( ) Future program managers are connectors.
( ) Programs focus on tactical deliverables achievement.
( ) Program managers spend more time on reporting.

7 What is the leadership example highlighted by Tommy Boy?
( ) There is only one view to leadership.
( ) Leadership is a discovery exercise.
( ) Leaders are born with it.
( ) It is easy to follow a predecessor in leading an organization.

Section II

Essential Skills to Lead Experience-Driven Cultures

Section Overview

This section gets to the heart of the matter when it comes to experiencing the program and project era that dominates tomorrow's cultures. Focusing on the project management muscle building and the implications this requires leaders to emphasize is a critical building block. The delicate balance between the human and the digital elements that shape the critical experiencing skills will have to be achieved. A strategic choice will be addressed pertaining to innovating in the next design wave of ways of working.

Section Learnings

- The relationships between human and digital capabilities, and how the balance is achieved to drive future cultures.
- The qualities of leading in the future in order to continuously create the most impact of project work.
- How do we achieve the continual adapting needed to thrive in tomorrow's organizations?
- Using the power of bridging and other enablers to shift approach and mindset in working across teams and stakeholders.
- Empowering the organizations' leaders to create inspiring models of working.

Creating Experience-Driven Organizational Culture: How to Drive Transformative Change with Project and Portfolio Management, First Edition. Al Zeitoun.

Keywords

- Muscles
- Inspiring
- Human
- Bridging
- Adapting
- Impact
- AI
- Balance

Introduction

The future cultures continue to be central for what will dominate the landscape of organizational design across organizations that want to sustain excellence and achieve larger societal commitments.

A few more of the eight driving hypotheses behind this work will be put to the test in this section to highlight how much shifting is expected in the toolbox of tomorrow's leaders of strategic initiatives. A few case studies will highlight the challenges that could be encountered if the right focus of leadership and management is not providing proper prioritization of what enables excellence across tomorrow's cultures. A few movies, such as *A Few Good Men*, will be utilized to establish linkages for how organizations could end up suffocating their employees and talent if the mission, transparency, and opportunity to do good meaningful work are lacking.

Some of my work with Kerzner, published in the PMWJ, will provide supporting details that flesh out the criticality of establishing a few of these human and digital skills shifts faster and more effectively. As a future leader, you will find that you must operate as a top executive without paying so much attention to rank, as social leadership is what will matter most going into the future of work. As the leader of a portfolio, you will have to be capable of depicting where the balance point should be established in relying on artificial intelligence (AI) as a complementing weapon while investing mostly in growing your emoting and bridging capabilities.

4

Project Impact Muscle

This chapter is focused on skills, namely project management skills, that are needed to create the proper experience-driven cultures of tomorrow. With the shifts that have been taking place to increase the scaling effects of transformation and strategic initiatives, it is becoming clearer that projects, or initiatives at large, are considered the core of how we work in the future.

Developing a project impact muscle has become a strategic priority. This requires a full commitment on the part of organizations and their management. This muscle building is a classic transformation exercise that combines people, process, and technology. On the people side, it requires a rethinking of what skills are critical and how much shifting to new models of leading is necessary. On the process side, it requires reimagining what must still remain and how much could be handled with more autonomy. On the technology side, there is a growing potential for a much-heightened focus on enhancing data strategies, how they become more central to running and changing the business, and thus having cross consequences on both the process and people slides.

Key Learnings

- Understand the value of reimagining the future of work in a way that supports how we do work in that future.
- Explore a case study that looks at the critical project impact on organizational excellence.
- Learn from a movie example how to develop the critical qualities necessary to drive transformational impact on teams.
- Address another of this work's hypotheses that confirm the value of projects and programs in driving growth.

Creating Experience-Driven Organizational Culture: How to Drive Transformative Change with Project and Portfolio Management, First Edition. Al Zeitoun.

4.1 The Future of Work

One of the critical changes ahead of organizations is the commitment to scaling transformational innovations, not only to drive growth but also to address the world's aspirations for a better future for the generations to come. This is another holistic capability for how to best drive the execution of strategy across the experience-driven cultures. Projects are being discovered as the most natural places for the experiencing, and they are also fulfilling the needs of motivated future generations.

I studied the next hypothesis in this work, related to this future of work, namely: ***Projects and programs are the standard vehicles for driving innovation and growth.***

The question I used was: "In your view, what contributes most to scaling innovation and growth?" Figure 4.1 shows the outcome of the polling related to this question. Dedicated projects and programs came second, which in my opinion was a good outcome of this polling. Projects and programs are the vehicles to execute the change needed for innovation. The key, and thus the highest score, is bold leadership. Leadership drives the shifts needed for the new view of the future of work, drives the building of the project impact muscle, and without any double, bold leadership allows us the space for the higher risk appetite necessary to take on some of the choices necessary for transformational innovation.

So, what does it take to develop bold leadership, and what could be preventing this from being a reality? The reason that bold leadership is quite relevant to the experience-driven cultures of the future is that these leaders are typically open for feedback, focused on innovation, and continuously come up with ideas that might

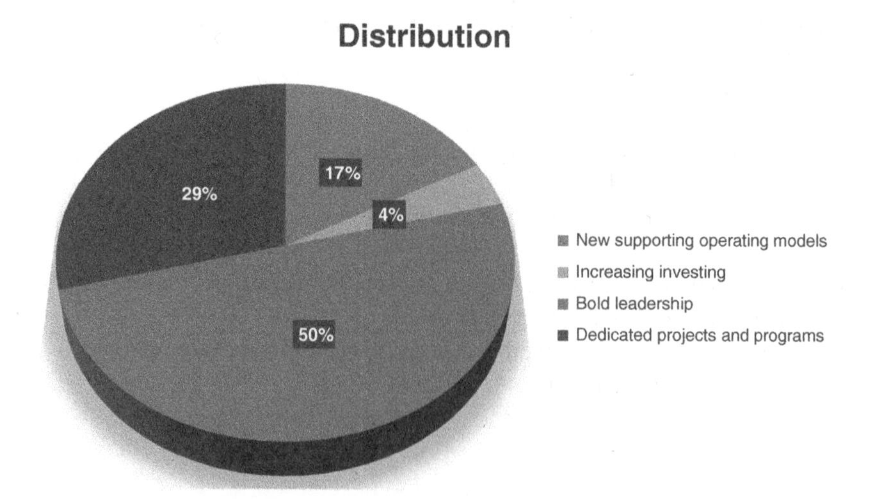

Figure 4.1 Scaling Innovation and Growth. *Note: Based on LinkedIn Open Polling, April 2024.*

be seen as extreme or controversial. Risk appetite and the strength of risk-taking are typically at the core of this type of leadership.

According to a 2016 Business Confidence Report by Deloitte,[1] bold leaders build breakthrough performance. Many key attributes were highlighted in that report. The types of goals that are set by these leaders tend to be ambitious. The willingness to get open feedback from many views and angles is another key attribute. Bold leaders tend to continuously look for new and improved ways of doing things. In addition, there are signs of boldness in their application of insane courage, what ideas they propose, the amount of risk they take, and most relevant to this work, the ways by which they empower their people toward strengthening the projects and programs teams and cascading this type of leadership into the teams.

Tip

Invest in developing a future of work that is powered by bold leadership. This improves the potential for projects and programs to scale innovation and growth.

4.2 Leading Projects

Leading projects in the future that are driven by artificial intelligence (AI) is a fresh new beginning. The future could be unsettling for the ones accustomed to leading projects in the classical ways. It is time to reset, and as seen in Figure 4.2, the future will see a collapse in the lines of demarcation between the human and the machine. The future of leading projects will have to be a mix of both humans and machines. It gives humans the opportunity to mature into a new version of themselves. What will remain irreplaceable is the emotional intelligence that humans bring.

Machines can help us with the unlimited potential of data and the enhancement of how we reach effective decisions. This means that when we lead in the future, we have to shift away from pure operational and reporting tasks to a much more strategic role. AI could enhance our risk management capabilities so we become much more proactive, which was the idea behind the true value of project management when it was envisioned as a profession, shifting away from the classical firefighting.

Leading project into the future, powered by AI, is a new dawn for the project profession. It will require a recipe that has adaptability at its center though. We could envision a future when the mundane tasks, the tactical management of schedule, budget, and resources toward achieving a given scope could all be automated. This leaves the door to crystallize the human value.

1 https://www2.deloitte.com/us/en/pages/operations/articles/cxo-confidence-survey.html.

Figure 4.2 Future of Projects. Credit: Vilkasss/Pixabay.

The true interactions with teams, the emoting, and the ability to thrive in being creative will be foundational in building the new forms of project leading ahead of us. What will the words used to describe the project manager be in the future? Who knows? Does the title really matter? Will there be a true management component in the role? Has that ever been truly the case? The new leader will master the handling of change in an increasingly uncertain world.

The following case will help address key elements of leading projects that are critical now and a few that will remain valuable in the future. The case will highlight where the human side of the equation could truly mature if automation is used to take care of implications of data and input changes. This would get organizations to a clarity level previously not achievable.

Case Study

Clark Faucet Company[2]

Background

By 2010, Clark Faucet Company had grown into the third-largest supplier of faucets for both commercial and home use. Competition was fierce. Consumers would evaluate faucets based on artistic design and quality. Each faucet had

2 Kerzner, 2022/John Wiley & Sons.

to be available in at least 25 different colors. Commercial buyers seemed more interested in the cost than the average consumer, who viewed the faucet as an object of art, irrespective of price.

Clark Faucet Company did not spend a great deal of money advertising on the radio, television, or Internet. Some money was allocated for ads in professional journals. Most of Clark's advertising and marketing funds were allocated to the two semiannual home and garden trade shows and the annual builders' trade show. One large builder could purchase more than 5,000 components for the furnishing of one newly constructed hotel or one apartment complex. Missing an opportunity to display the new products at these trade shows could easily result in a 6- to 12-month window of lost revenue.

Culture

Clark Faucet had a noncooperative culture. Marketing and engineering would never talk to one another. Engineering wanted the freedom to design new products, whereas marketing wanted final approval to make sure that what was designed could be sold.

The conflict between marketing and engineering became so fierce that early attempts to implement project management failed. Nobody wanted to be the project manager. Functional team members refused to attend team meetings and spent most of their time working on their own pet projects rather than on the required work. Their line managers also showed little interest in supporting project management.

Project management became so disliked that the procurement manager refused to assign any of his employees to project teams. Instead, he mandated that all project work come through him. He/she eventually built a virtual brick wall around his employees. He/she claimed that this would protect them from the continuous conflicts between engineering and marketing.

The Executive Decision

The executive council mandated that another attempt to implement good project management practices must occur quickly. Project management would be needed not only for new product development but also for specialty products and enhancements. The vice presidents for marketing and engineering reluctantly agreed to try to patch up their differences but did not appear confident that any changes would take place.

(Continued)

Case Study (Continued)

Strange as it may seem, no one could identify the initial cause of the conflicts or how the trouble actually began. Senior management hired an external consultant to identify the problems, provide recommendations and alternatives, and act as a mediator. The consultant's process would have to begin with interviews.

Engineering Interviews

The following comments were made during engineering interviews:

- "We are loaded down with work. If marketing would stay out of engineering, we could get our job done."
- "Marketing doesn't understand that there's more work for us to do other than just new product development."
- "Marketing personnel should spend their time at the country club and in bar rooms. This will allow us in engineering to finish our work uninterrupted!"
- "Marketing expects everyone in engineering to stop what they are doing in order to put out marketing fires. I believe that most of the time the problem is that marketing doesn't know what they want up front. This leads to change after change. Why can't we get a good definition at the beginning of each project?"

Marketing Interviews

These comments were made during marketing interviews:

- "Our livelihood rests on income generated from trade shows. Since new product development is four to six months in duration, we have to beat up on engineering to make sure that our marketing schedules are met. Why can't engineering understand the importance of these trade shows?"
- "Because of the time required to develop new products [four–six months], we sometimes have to rush into projects without having a good definition of what is required. When a customer at a trade show gives us an idea for a new product, we rush to get the project underway for introduction at the next trade show. We then go back to the customer and ask for more clarification and/or specifications. Sometimes we must work with the customer for months to get the information we

need. I know that this is a problem for engineering, but it cannot be helped."

The consultant wrestled with the comments but was still somewhat perplexed. "Why doesn't engineering understand marketing's problems?" pondered the consultant. In a follow-up interview with an engineering manager, the following comment was made: "We are currently working on 375 different projects in engineering, and that includes those that marketing requested. Why can't marketing understand our problems?"

Questions

1. What is the critical issue?
2. What can be done about it?
3. Can excellence in project management still be achieved and, if so, how? What steps would you recommend?
4. Given the current noncooperative culture, how long will it take to achieve a good cooperative project management culture and even excellence?
5. What obstacles exist in getting marketing and engineering to agree to a single methodology for project management?
6. What might happen if benchmarking studies indicate that either marketing or engineering is at fault?
7. Should a single methodology for project management have a process for the prioritization of projects, or should some committee external to the methodology accomplish this?

Reflections: Leading in the future will require deeper listening to what the true causes of a challenge might be. This will have to be coupled with creative problem solving that is anchored in better sensing of the gaps toward achieving results. The case study also highlighted the need for building a culture of collaboration. This supports the experiencing that makes it easier for departments and senior leaders, with different agendas, to better see others' points of view. A true simulation of what success would look like could be created.

Tip
Portfolio management remains a critical enabler for leading into the future. Leaders need to have a unified view of how they prioritize the work to meet the most critical choices.

4.3 Impact Matters

Kerzner and Zeitoun (2022) address the rethinking of the role of the project manager to drive the creation of impact. In the cultures of the future, the role has to shift to create the differentiating impact.

4.3.1 Introduction

When we look at the word project management, our minds can still take us to a place where rigor, structure, and control processes prevail. The last decade has altered much of our conviction about what great project management is. We learned a lot about what no longer works, the true definitions of what good looks like, and the qualities of the person to whom we might still give the title of project manager. The next decade is the most critical test for this profession, and many authors and practitioners admire and believe in its impact.

This look ahead requires us to build on the learnings of this past decade. The unprecedented reliance on digitization, the intense collaborative working from every possible corner of the universe, the shift in the ways of working and frameworks, and the understanding of the impact of projects and programs on creating change have opened a new page for experimenting with the role of the project manager. Is that role truly about managing, is it about leading, is it about both, or is it an emerging set of ingredients that should be categorized differently and given new names? The evidence of organizational excellence continues to center on practices that agree to a set of principles, focus on execution, use a higher trust currency, and realize the unlimited potential of projects in this project economy for making sustainable changes stick.

The reinvention of the project manager is upon us and requires a degree of commitment to rebuilding the future organization to be the strategic, innovative, and learning community it will have to become for meaningful and sustainable strategic successes to prevail.

4.3.2 Forecasting Changes to the Role of the Project Manager

When we combine the words "forecasting" and "project management," we envision a process of making predictions or assumptions about the possible outcomes of a project. We perform an analysis of historical project data as well as guesses on future outcomes to determine the duration, cost, and performance at project completion.

Project forecasting is done continuously and on every project. What companies fail to do is to forecast what the role of the project manager will be in the future based on the major changes identified in the project management community of practice or changing roles within the organization.

Past success in project management is no guarantee of future performance. The management guru, Peter Drucker, often used the term, "The Failure of Success," where companies become so successful at what they are doing that they refuse to challenge the results and the accompanying processes to see if it can be accomplished better in the future.

4.3.3 Drivers for Role Change

Forecasting major changes to the role of the project manager must begin with an understanding of the drivers that will necessitate the changes expected to take place.

Some of the drivers for role changes include the need for the project manager to:

- Manage new types of projects.
- Design and select new types of methodologies for new types of projects.
- Make business as well as technical decisions.
- Select new types of metrics for the new decisions required.
- Use information warehouses, business intelligence systems, and digital technologies.

Developing new types of project leadership skills and being able to collaborate with all stakeholders more effectively will shape the DNA of the future project manager. One of the biggest mistakes executives still make is limiting their views of what project managers are capable of achieving for their organizations strategically. The traditional tactical view of projects' value has shifted to strategic value in most of the world's organizations that have exemplified a pattern of consistent growth. These organizations now holistically measure what matters and thus understand the shifts in the role of the project managers that get them there.

4.3.4 Types of Projects

Project managers have been managing traditional projects for decades. Traditional projects have well-defined requirements, a business case, a statement of work, and possibly a complete work breakdown structure for all the work broken down into five or six levels of detail. Project managers are now being asked to manage strategic projects, such as innovation and research and development (R&D), that begin with just an idea, and the scope of the effort is progressively elaborated as the work takes place.

New types of projects usually require new types of leadership and new types of decisions to be made. On traditional projects, the governance committee or project sponsor often had a major role in making business decisions. On strategic projects, business decision-making is becoming a project management responsibility.

4.3.5 Making Business Decisions

In the early years of modern project management practices, project management evolved from the aerospace, defense, and heavy construction industries. Most of the project managers were engineers who were assigned to the projects because of their command of technology. The criteria for being assigned as a project manager were a command or good understanding of technology accompanied by writing skills. Business-related decisions were most often made by governance personnel and project sponsors.

Senior management over the years has realized that, as the number of projects has increased, executives do not have the time to act as sponsors on all projects. Allowing PMs to make business decisions meant that senior management had to rethink whether a command-and-control leadership model from the top floor of the building was the best approach. Senior management surrendered the idea that information is power and began sharing strategic information with project teams.

Today, there exists a line-of-sight between project teams and senior management to make sure that all projects are aligned to strategic business objectives. Knowledge of strategic business objectives is a necessity if PMs are expected to make business decisions and interface with stakeholders.

Many of the decisions made by project managers on traditional projects were heavily focused on short-term profitability and short-term decisions. The management of strategic projects focuses on decision-making, affecting long-term rather than short-term expectations.

4.3.6 The Fuzzy Front End (FFE)

Companies are now rethinking when to bring project managers on board the project. Project selection and prioritization is referred to as the FFE. Historically, senior management selected the projects, assigned a priority to the projects, and then assigned a project manager responsible for project execution.

The problem with this approach was that the PMs had a poor understanding of how the executives selected the projects, the factors they considered in selection and prioritization, the risks they considered, the business benefits and value they expected, and most often the budget and schedule provided were insufficient. In the future, we can expect project managers to be brought on board during the FFE to assess the resources needed, whether the technology needed is available, and whether the expectations are realistic.

4.3.7 New Metrics

Perhaps the most significant change that will take place will be the use of new metrics. When project managers are expected to make only technical decisions, the metrics of time, cost, and scope that are included in the earned value measurement system may be sufficient. However, if project managers are expected to make business and strategic decisions, then significantly more metrics will be required. Some of the new metrics that project teams will require include:

- Metrics that track the creation of business benefits and business value.
- Metrics that measure intangibles such as the effectiveness of project governance and customer satisfaction.
- Metrics that measure strategic issues related to the project such as how well the project is aligned with strategic business objectives.

Another category of metrics that is growing includes metrics related to risks. Traditional metrics report progress and issues but usually not the cause of problems, especially potential problems or risks that can lead to failure. Failure does not occur at the end of a project. There are always indicators or metrics that, if used as part of monitoring and control right from the start of the project, could provide an early indication that a potential risky situation is about to occur. This could allow teams to correct risky situations early in the project's life cycle. These metrics can serve as an early warning system. Unfortunately, they are not part of traditional Earned Value Management System (EVMS) usage. Some of these critical metrics include:

- The number of new assumptions made over the project's life cycle.
- The number of assumptions that changed over the project's life cycle.
- Changes that occurred in the enterprise due to environmental factors.
- The number of scope changes approved and denied.
- The number of time, cost, and scope baseline revisions.
- The effectiveness of project governance.
- Changes in the risk level of the critical work packages.

4.3.8 Methodologies

The days of using a one-size-fits-all methodology are disappearing. The new types of projects and the new decisions that project managers will be making will necessitate giving project teams a choice of flexible methodologies to use. At the onset of a project, the team will select the best methodology for the project. In an ideal situation, the methodology selected, as well as the life cycle phases, will be aligned with the customer's business model if possible. This will build customer satisfaction and trust, accelerate the decision-making process, and generate repeat business. This can make life much easier for customers to track the project and provide the correct and timely support when needed.

The team will also select the metrics they need for the decisions they make and the information requested by the stakeholders. Each stakeholder may have different information needs. A dashboard designer will be assigned to each project team to customize the dashboards that stakeholders request.

4.3.9 Leadership

Over the years, project management leadership has focused heavily on the use of authority and power. Attempts were made to adapt traditional functional leadership models into a project management environment, and many attempts were unsuccessful. A new form of leadership is emerging, namely social project management leadership, which focuses on ways to get team members more engaged in the project. Some of the factors driving new forms of project management leadership include:

- New types of projects are requiring a greater need for collaboration with team members and stakeholders.
- Projects are becoming longer in duration, and project managers have more time to interface with team members and understand their needs.
- Project managers are increasingly providing input into team members' performance reviews.

Project management leadership decisions will focus on business as well as technical decisions.

4.3.10 Change Management

The outcome of many projects requires changes to be made in how the company conducts its business. Projects will be needed to align the deliverables with business growth needs. In the past, project managers did not have an active role in change management practices. In the future, project managers can be expected to take the lead in implementing the changes needed because of the project's deliverables.

4.3.11 Crisis Management

In most companies, crisis committees are chaired by senior management and accompanied by a command-and-control leadership style from senior management. The COVID-19 pandemic resulted in most people working from home and made it clear that the role of the project manager was increasing in importance. Project managers were required to manage projects using virtual teams and find ways to determine the mental health of the team members. New project management leadership styles will be needed to engage team members using virtual meeting technologies.

4.3.12 The Path Forward

The changing role of the project manager, with a heavy focus on business strategy, is forcing project managers to develop many new skills, especially business-related skills. These skills will be transferable to other job opportunities.

The good news here is that executives are gradually comprehending the strategic potential of project management and thus paying closer attention to the leaders running the initiatives. The impact of project managers is finally reaching a level of clarity not seen before. As an example, even though some organizations have used titles like Chief Projects Office over a couple of decades ago, they did not fully understand till recently the true nature of that executive position and how much it could contribute to transforming every aspect of businesses and their impact on key stakeholders. No one is better equipped in the future to lead transformation than properly and strategically prepared project managers.

Shifts in how we prioritize and do work, how we run dialogues and execute programs and projects, and how we integrate these efforts with the right metrics and new views of success will shape much of the role changes ahead. It is our sincere wish that this reinvention of the project manager will have a lasting impact on a globe that is dealing with more uncertainty than at any time in recent history and is affected by complex disruptions that are much more difficult to predict. It is with adaptability, resilience, and true belief in the diverse views of talented project and program team members that we will be able to achieve the shifts in organizational and governmental leadership that will transform the project management skills to the level of achieving missions and outcomes that matter.

4.4 Building Project Muscles

Building and growing the project muscles has become a strategic priority. It is the recipe for tackling what it takes to design the future of work, lead in the future, and create the impact parts of this chapter addressed above. In this new dawn that is AI-powered, the human dimension for driving focus and motivation is of utmost value. Creating the right experiences for the team is a key role for the project manager in building project muscles for the future.

Let's take the example of another movie to illustrate the development of these muscles. In the movie, "**Any Given Sunday**," Al Pacino, the head coach of the Miami Sharks, exhibits some of the critical attributes to building these project muscles as he tries to take his struggling team back to safer harbors.

When the team struggled with consecutive losses, diminishing interest in game attendance, an older quarterback, Jack, A; Pacino had to reinvent himself in order to balance this team struggle, the new expectations of the young owner of the team, and his own personal challenges. He had to step up and do what is right to reignite the team's soul and motivation, regardless of the arrogance and bad examples set by some of the other younger and possibly effective players who were not helping with creating a unified team experience. Al Pacino followed a path similar to what is shown in Figure 4.3 to reinform the team and get the excitement back for the diverse set of stakeholders.

Just like in the example of this movie, it is critical that leaders in the future, regardless of how much we will be able to create use of responsible and ethical AI, will have to be life learners. This will be needed to create the competencies for creative problem solving, inspiring, leading through complexity, better handling of unknowns, and building connected fabric for teams and across the organizations.

> **Tip**
> Adapting to the changing dynamics of a project is core to the muscles to develop. Building project muscles is anchored in growth mindset and nonstop learning.

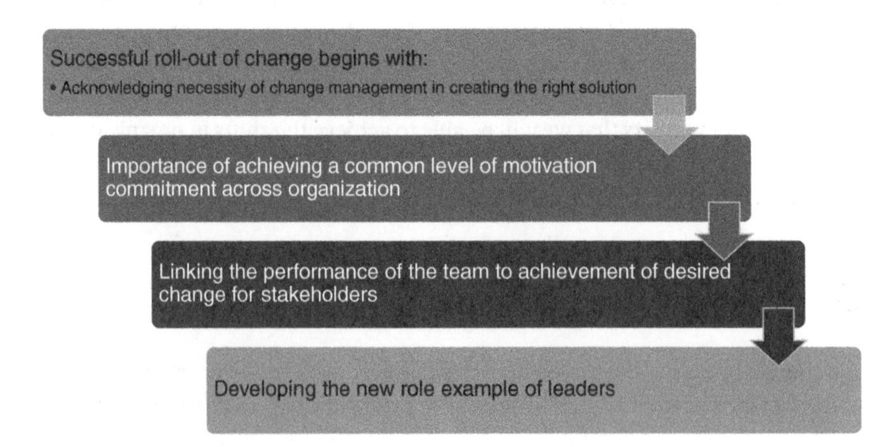

Figure 4.3 Building Project Muscles.

Reference

Kerzner, H. and Zeitoun, A. (2022). The Reinvention of the Project Manager: The Great Project Management Accelerator Series, *PM World Journal*, Vol. XI, Issue XI.

Review Questions

Parentheses ( ) are used for Multiple Choice when one answer is correct. Brackets [ ] are used for Multiple Answers when many answers are correct.

1 Which of the following is a true differentiator in the future of work that directly supports innovation and growth?
   ( ) Increased level of investing.
   ( ) Bold leadership.
   ( ) Full reliance on AI.
   ( ) A framework that has to be applied across all innovation initiatives.

2 What are good ingredients for bold leadership? Choose all that apply.
   [ ] Ideating regardless of how ideas might seem.
   [ ] Being an executive.
   [ ] Ambitious goals.
   [ ] Higher appetite for risk-taking.

3 What is the greatest value of having AI-powered future projects leading?
   ( ) Ensures that all decisions are automated.
   ( ) Depend on data for all reporting requirements.
   ( ) Making project managers focus on strategic value.
   ( ) Enabling organizations to restructure project managers.

4 What is the future value of the title project manager?
   ( ) It adds prestige to the role of the manager.
   ( ) It is an opportunity to highlight what the leader does.
   ( ) Title will matter less, and what counts will be the reinvented role of the project manager.
   ( ) Showing off the leader's capabilities.

5 What were the core challenges in the Clark Faucet Company's ways of working? Choose all that apply.
   [ ] Engineering could not understand where marketing is coming from.
   [ ] Project management was well respected.
   [ ] Marketing could not understand what the engineering problem is.
   [ ] Focus on getting one project management methodology utilized.

(Continued)

(Continued)

6 What is the key focus of creating impact in projects?
( ) Keep the role of the project manager relevant.
( ) Depends on the nature of the project and success definition.
( ) Speed of getting things done.
( ) Spending less time on reporting.

7 What is the building project muscles example highlighted by Al Pacino?
( ) There is only one way to success in building a team.
( ) Building muscles is an experiencing exercise.
( ) Young players matter most.
( ) It is easy to create a connected team.

5

Effective Experiencing

Effective experiencing is easy to spot. One could see that in the energy of the organization and how much an organization lives up to its ethos. Organizations that are clear about their distinct cultural characteristics are committed to creating and sustaining effective experiences for their stakeholders.

This becomes a top priority for the management team and becomes a natural daily commitment for every player in the extended ecosystem. The words to describe such organization could be innovative, learning, empowering, and growth oriented.

Creating effective experiences is both an art and a science. The art consists of the human elements, especially around reinventing how we lead, guide, coach, and create the well-connected thread among all key stakeholders. The science covers the techniques, technology, and most importantly the data that enables the iterative nature of work that is expected in the experience-driven world of tomorrow.

A powerful formula surfaces that balances both elements of the formula and enables leaders to continuously mature their strategic and operational work to focus on value creation.

Key Learnings

- Understand the features that are the clear signs of an experience-driven culture.
- Learn how to continually adapt and build the muscle for effective change leadership.
- Explore the principles of social leadership and how it contributes to building effective teams.
- Understand the potential of shifting the culture and execution practice with a new mindset.
- Develop the momentum for the engine behind sustaining a creative organization.

Creating Experience-Driven Organizational Culture: How to Drive Transformative Change with Project and Portfolio Management, First Edition. Al Zeitoun.

5.1 Experience-Driven Features

To construct the features of an experience-driven environment, many brains need to come together and invest in shaping what that successful view of the future would look like. As shown in Figure 5.1, multiple input points should come together to create the most effective description of those features. This way the outcome would be something that the entire team and management would stand behind and commit to.

As we brainstorm the features that matter, we got to make space to think and ensure that the basic enablers to support teamwork, are in place. Agreeing on the right features by itself is a great step in designing an experience-driven environment. It confirms that we are able to use research, data, and ideas well. We would then be in a position to combine, evaluate, and make the recommended decision on solutions as the key words around the light bulb indicate.

The following case study is an ideal example to set the stage for this chapter and the multiple steps that need to be foundationally created for an effective experience outcome. Lego with its beautiful colorful plastic pieces drives our innovation to construct, create, and solve challenges. It is loved by many across genders and generations. The case study will serve many of the learning points that need to be highlighted and understood to ignite effective experiencing.

Figure 5.1 Features Development.

Case Study

Lego: Brand Management[1]

Abstract

Lego is one of the most admired companies in the world. Yet despite their success, they went through a period that put them on the brink of bankruptcy. They eventually changed their corporate culture and reconfigured their product development and innovation processes to turn the company around.

Although most of the issues discussed in the case can occur in any company, the Lego case illustrates the challenges facing an extraordinarily successful privately held company where innovation was needed to support the growth of the Lego brand.

Portions of this case study have been adapted from Wikipedia, the Free Encyclopedia: Lego; the Lego Group; Lego Minifigures; and Lego Mindstorms. The first part of the case study focuses on the products and services provided by Lego, not necessarily in chronological order, so that the reader can understand the interactions with various forms of brand innovation and why cultural changes were necessary. The focus of the case is to provide an understanding of managing projects under a brand, not the effectiveness or ineffectiveness of managerial decisions.

Understanding Brand Management

One of the most difficult types of innovation projects to manage are those that must support brand management activities. A brand could be the company, a product or family of products, specific services, or people. Successful brands may take years to develop and continuous innovation is necessary to maintain brand awareness, credibility, and consumer loyalty. Companies with strong brand awareness include Apple, Google, Disney, Microsoft, Coca-Cola, Facebook, and Lego.

Brand management practices are heavily oriented around marketing activities that focus on the target markets and how the product or family of products look, are priced out, and are packaged. Brand management must also focus on the intangible properties of the brand and the perceived value to the customers. An intangible property might be the value your customers place

(Continued)

1 Kerzner, 2022/John Wiley & Sons.

Case Study (Continued)

on your products such that they are willing to pay more than the cost of a generic brand that may function the same.

Almost all brand innovations involve governance by the brand manager whose responsibility is to oversee the relationships that the consumers have with the brand, thus increasing the value of the brand over time. Innovations should allow for the awareness and pricing of the brand to grow as well as maintain or improve consumer brand loyalty.

Unlike other forms of innovation where project teams have the freedom to explore multiple options and ideas and go off on tangents, brand innovation may have restrictions established by brand management. Brand management is responsible not only for managing and promoting the brand but also for deciding what new products or innovations could fall under the brand umbrella. Brand innovation is a marriage between the brand's core values, the target market, and management's vision. The difference between the long-term success and the failure of a brand rests with brand innovation.

History

Lego, which conducts business as the Lego Group, is a privately held toy manufacturing company headquartered in Billund, Denmark. It is best known for its flagship product, the colorful interlocking plastic bricks accompanied by an array of gears, figurines called minifigures, and various other parts. There are more than 7000 different Lego elements. Lego pieces can be assembled and connected in many ways to construct objects, including vehicles, buildings, and working robots. Anything constructed can be taken apart again, and the pieces reused to make new things. As of July 2015, 600 billion Lego parts had been produced.

The history of Lego spans nearly 100 years, beginning with the creation of small wooden playthings during the early 20th century. Manufacturing of plastic Lego bricks began in Denmark in 1947 and has since grown to include factories throughout the world. Movies, games, competitions, stores, and Legoland amusement parks have been developed under the Lego brand. Lego has more than 40 global offices. By the turn of the century, Lego was producing more than 20 billion Lego bricks a year and was functioning as both a retailer and an entertainer.

The company was founded on August 10, 1932, by Ole Kirk Christiansen (1891–1958). The brand name "Lego" is derived from the Danish words "leg godt," meaning "play well." The Lego Group's motto is "Det bedste er ikke for godt," which means "Only the best is good enough." This motto, which is still used today, was created by its founder to encourage his employees never to skimp on quality, a value he believed in strongly. By 1951 plastic toys accounted for half of the Lego company's output, even though the Danish trade magazine Legetøjs-Tidende ("Toy Times"), visiting the Lego factory in Billund in the early 1950s, felt that plastic would never be able to replace traditional wooden toys. Lego toys seem to have become a significant exception to the common sentiment of expressing dislike for plastic in children's toys.

Lego Bricks

By 1954, Christiansen's son, Godtfred, had become the junior managing director of the Lego Group. His conversation with an overseas buyer led to the idea of a toy system. Godtfred saw the immense potential in Lego bricks to become a system for creative play, but the bricks still had some problems from a technical standpoint; their locking ability had limitations and lacked versatility. In 1958, the modern brick design was developed. It took five years to find the right material for it, which was a polymer called acrylonitrile butadiene styrene (ABS). The modern Lego brick design was patented on January 28, 1958.

Lego pieces of all varieties constitute a universal system. Despite variations in the design and the purposes of individual pieces over the years, each piece remains compatible in some way with other pieces. Lego bricks from 1958 still interlock with those made today, and Lego sets for young children are compatible with those made for teenagers. Six bricks containing 2×4 studs can be combined in 915,103,765 ways.

Each Lego piece must be manufactured to an exacting degree of precision. When two pieces are engaged, they must fit firmly, yet be easily disassembled. The machines that manufacture Lego bricks have tolerances as small as 10 µm.

Manufacturing

Manufacturing of Lego bricks occurs at several locations around the world. Molding is done in Billund, Denmark; Nyíregyháza, Hungary; Monterrey, Mexico; and most recently in Jiaxing, China. Brick decorations and packaging are done at plants in Denmark, Hungary, Mexico, and Kladno in the Czech Republic. The Lego Group estimates that in the last five decades, it has

(Continued)

Case Study (Continued)

produced 400 billion Lego bricks. Annual production of Lego bricks averages approximately 36 billion, or about 1140 elements per second. According to an article in Business Week in 2006, Lego could be considered the world's number one tire manufacturer because the factory produces about 306 million small rubber tires a year. The claim was reiterated in 2012.

Lego's Target Consumers

Lego's target consumers were originally boys aged four to nine, although 10–20% of their total customers were girls. Their consumers were usually members of families that wanted their children to grow up as scientists, architects, designers, and even musicians.

To connect the Lego brand to consumers, Lego had to conduct research and work closely with families to understand how children play and spend their time. In some countries, Lego discovered that parents wanted toys that children could play with by themselves, without supervision, whereas in other countries, parents wanted to sit on the floor and accompany their children playing with the toys. Most families have rules that children must abide by concerning how many hours a day they can watch television or play on computers. Although there were several options available for entertainment, Lego believed that children wanted the freedom to show their creativity with the plastic bricks by building something masterful. They could use their imagination to build something to be proud of, and then make up stories using the Lego action figures.

As computers and wireless broadband technology expanded into the children's bedrooms, Lego had to rely upon radical innovation and develop new products to expand their services into the virtual space, while remembering that construction rather than technology drives Lego's business. This included videos on large Lego construction projects, video games, board games, movies, and the ability to share experiences or play games with other Lego users. The goal was for consumers to continue purchasing licensed Lego products. Adults were also purchasers of Lego products, giving them the chance to relive their childhood. One adult spent two years building a Lego playable harpsichord made with 100,000 bricks.

Lego also developed a strategy for school teachers who taught the targeted age groups by developing hands-on kits for teachers. Lego products were recognized as a compromise between imagination, creativity, and fun for children. However, purchasing the kits was limited by the school's budget. In 2015, an article was published stating that children with autism were improving their long-term social interaction skills using Lego products.[2]

In 1998, Lego suffered its first financial loss. The following year, Lego signed a licensing agreement with Lucasfilm to create Lego kits for the Star Wars movie series. This was a shift in Lego's innovation strategy from just open-ended play kits to branded kits based upon movie themes. It was also a departure from Lego's desire not to make "war toys." Lego's gross sales jumped 30% in the first year and sales of the Star Wars kits exceeded expectations by 500%. Many of the buyers were adults who purchased the kits for nostalgia reasons.

Lego soon recognized the benefits of targeting adults as well as children. Adults were willing to pay $800 for the Star Wars Millennium Falcon Kit, $500 for the Star Wars Death Star, and $400 for Harry Potter's Hogwarts Castle. However, there were some issues. Kids were willing to let their imagination run wild, pretend they were part of the theme, and then quit and disassemble the pieces. Adults needed foolproof assembly instructions because the satisfaction of the finished product was what motivated them. Some adults even posted time-lapse videos on YouTube showing how they constructed large Lego buildings and other products. Another benefit for adults to purchase Lego products was to reduce stress.

Many companies had a room with Lego products where employees could reduce stress, meditate, relax, and drown out noise as they undertook a creative challenge. Lego hired Abbie Headon to write a book focusing on adult usage of Lego products, entitled Build Yourself Happy: The Joy of Lego Play.

Gender Equality

By the mid-to-late 1980s, Lego almost trapped themselves with the "failure of success" by not realizing that their market was changing. Most of Lego's revenue appeared to be coming from boys; girls were losing interest in Lego's building blocks. In discussions with psychologists, Lego discovered that girls were developing interests other than toys at a much earlier age than boys. However, this was dependent on the area of the world someone lived in. Lego also discovered that girls preferred the pastel colors while boys preferred the sharper colors of black, blue, red, yellow, and green.

(Continued)

2 Barakova, Emilia I.; Bajracharya, Prina; Willemsen, Marije; Lourens, Tino; and Huskens, Bibi. Long-term LEGO therapy with humanoid robot for children with ASD. Expert Systems 32 (6, December 2015) 698–709.

Case Study (Continued)

In 2003, Lego launched Clikits, a product designed specifically for girls six years old and older. It contained arts-and-crafts materials from which girls could design jewelry, accessories for hair and fashion statements, and picture frames. For boys, Lego developed a series of table sports products where they could simulate playing basketball, baseball, and hockey.

In 2012, Lego Friends was launched, which was a collection of Lego construction toys designed primarily for girls. The theme introduced the "mini-doll" figures, which were about the same size as the traditional minifigures but were more detailed and realistic. The female mini-doll figures could only sit, stand, or bend over, whereas male minifigures used for games for boys had more flexibility and could drive cars, run, and hold tools.

The Lego Friends sets include pieces in many color schemes such as orange and green or pink and purple and depict scenes from suburban life set in the fictional town of Heartlake City. The main characters, one of which appears in every set, are Andrea, Emma, Mia, Olivia, and Stephanie. The sets were usually named after them. In the initial wave of sets, the larger sets included bricks that could build a veterinary clinic, a malt-style café, a beauty salon, and a suburban house; smaller sets included a "cool convertible," a design studio, an inventor's workshop, and a swimming pool. The Friends product replaced previous female-oriented themes.

Following its launch, the girl-friendly Lego Friends was deemed offensive by many and considered one of the worst toys of the year. But at the same time, plenty of young girls and their families had a different opinion. Despite the criticism by some, Lego Friends was an impressive financial success and was honored by the Toy Industry Association as the Toy of the Year in 2013.

In January 2014, a handwritten letter to Lego from a seven-year-old American girl, Charlotte Benjamin, received widespread attention in the media. In it, the young author complained that there were "more Lego boy people and barely any Lego girls" and observed that "all the girls did was sit at home, go to the beach, shop, and they had no jobs, but the boys went on adventures, worked, saved people, and even swam with sharks."[3]

In June 2014, it was announced that Lego would be launching a new "Research Institute" collection of toys featuring female scientists, including

3 Gander, Kashmira. "Lego told off by 7-year-old girl for promoting gender stereotypes," The Inde- pendent, February 3, 2014. Alter, Charlotte, "Soon There Will Be Female Scientist Legos," Time, June 4, 2014.

a female chemist, paleontologist, and astronomer. The new sets showed women doing intellectually challenging jobs. Lego denied claims that the set was introduced to placate criticism of the company by activists. The Research Institute collections sold out within a week of its online release in August 2014.

Core Values

Over Lego's existence, the company has endured ups and downs in its business. As new products were developed to satisfy changing market conditions, Lego had to revise its business model. Business models also changed due to new partnerships and licensing agreements. But what had not changed in all that time was Lego's strategic vision and core values.

Lego's strategic objectives were to continue to be creative in developing new toys and to reach out to more children each year. The strategic objectives were supported by Lego's core values, which focused on the "pride of creation, high quality, a strong hands-on element, and fun." All products were designed around these core values. Most of the downturns in their business occurred when they deviated too far from their core values.

These core values were also critical when Lego selected licensing partners. As stated by Jorgen Vig Knudstrop, past CEO of the Lego Group, "It's important that the licensing partners we work with provide an experience that is on par with Lego's [core] values."

Licensing Agreements

The Lego Group's licensing agreements fall under two distinct categories: inbound and outbound. Inbound licensing refers to the licensees that grant Lego permission to create licensed themes from such movies (or movie series) as Star Wars, Winnie the Pooh, Batman, Indiana Jones, Lord of the Rings, and Marvel comics superheroes, including Spider-Man. Outbound licensing refers to examples where a company is given permission to use the Lego Group's intellectual property, such as for publishing books or for Merlin Entertainments' use in Legoland theme parks. Outbound licensing also includes apparel, luggage, lunch boxes, electronics, school supplies, media games, clocks, and watches.

Lego's Legacy of Success

Lego has maintained a legacy of success over its life. Lego's popularity is demonstrated by its wide representation and usage in many forms of cultural

(Continued)

Case Study (Continued)

works, including books, films, and artwork. Other than the traditional bricks, Lego-branded products include apparel, footwear, backpacks, party goods, greeting cards, children's bedding, Halloween costumes, watches, oral care, board games, and publishing.

In 1998, Lego bricks were one of the original inductees into the National Toy Hall of Fame in Rochester, New York. In 1999, Lego bricks was named the "Toy of the Century" by Fortune. In 2011, in a survey by the Research Institute, Lego was the number one admired brand in Europe, number two in the United States and Canada, and number five globally.[4] By the first half of 2015, the Lego Group became the world's largest toy company by revenue, with sales amounting to US$2.1 billion, surpassing Mattel, which had US$1.9 billion in sales. In February 2015, Lego replaced Ferrari as Brand Finance's "world's most powerful brand." In May 2018, the company made it to Forbes' 100 World's Most Valuable Brands 2018, at 97th on the list.

Trademarks and Patents

Since the expiration of the last standing Lego patent in 1989, several companies, including Tyco Toys, Mega Bloks, and Best-Lock, have produced interlocking bricks similar to Lego brand bricks. These competitors' products were typically compatible with Lego brand bricks and were often marketed at a lower cost than Lego sets.

One such competitor was the Chinese company Tianjin Coko Toy Co., Ltd. In 2002, the Lego Group's Swiss subsidiary, Interlego AG, sued the company for copyright infringement. A trial court found many Coko bricks to be infringing. Coko was ordered to cease manufacture of the infringing bricks, publish a formal apology in the Beijing Daily, and pay a small fee in damages to Interlego. On appeal, the Beijing High People's Court upheld the trial court's ruling.

In 2003, the Lego Group won a lawsuit in Norway against the marketing group Biltema for its sale of Coko products, on the grounds that the company used product confusion for marketing purposes. Also, in 2003, a large shipment of Lego-like products marketed under the name "Enlighten" was seized by Finland customs authorities. The packaging of the Enlighten products was like official Lego brand packaging. Their Chinese manufacturer failed to appear in court, and thus Lego won a default action ordering the destruction of the shipment. The Lego Group footed the bill for the disposal of the

4 Brad Wieners, "Lego Is for Girls," Bloomberg Businessweek, December 25, 2011.

54,000 sets, citing a desire to avoid brand confusion and protect consumers from potentially inferior products.

Not all patent and trademark lawsuits resulted in favor of the Lego Group. In 2004, Best-Lock Construction Toys defeated a patent challenge from Lego in Oberlandesgericht, Hamburg, Germany.

The Lego Group attempted to trademark the "Lego Indicia," the studded appearance of the Lego brick, hoping to stop production of Mega Bloks. On May 24, 2002, the Federal Court of Canada dismissed the case, asserting the design is functional and therefore ineligible for trademark protection. The Lego Group's appeal was dismissed by the Federal Court of Appeals on July 14, 2003. In October 2005, the Supreme Court ruled unanimously that "Trademark law should not be used to perpetuate monopoly rights enjoyed under now-expired patents" and held that Mega Bloks can continue to manufacture their bricks.

Because of fierce competition from copycat products, the company has always responded by being proactive in its patenting and has over 600 United States–granted design patents to its name.

Environmental Issues

Lego maintains a corporate social responsibility program that says that it was "our ambition to protect children's rights to live in a healthy environment, both now and in the future."[5] Lego was pressured by environmental groups to acknowledge the impact of its operations on the environment, especially in areas such as climate change, resource and energy use, and waste. All manufacturing sites were certified according to the environmental standard ISO 14001. Lego began seeking alternatives to crude oil as a raw material for its bricks. This resulted in the establishment in June 2015 of the Lego Sustainable Materials Center as a significant step toward the 2030 ambition of finding and implementing sustainable alternatives to current materials.

In 2011, Lego bowed to pressure from the environmental campaigning organization Greenpeace, reportedly agreeing to drop supplier Asia Pulp and Paper, and pledging to only use packaging material certified by the Forest Stewardship Council in future. The environmental group had accused Lego, Hasbro, Mattel, and Disney of using packaging material sourced from trees cleared out of the Indonesian rainforest.

(Continued)

5 Olsen, P. E., "Save the Whales? A Public Relations Crisis at Lego," Journal of Critical Incidents 8 (October 2015): 130.

Case Study (Continued)

Lego partnered with the oil company Royal Dutch Shell in the 1960s, using the company's logo in some of its construction sets. This partnership continued until the 1990s and was renewed again in 2011. In July 2014, Greenpeace launched a global campaign to persuade Lego to cease producing toys carrying the oil company Shell's logo in response to Shell's plans to drill for oil in the Arctic. By August 2014, more than 750,000 people worldwide had signed a Greenpeace petition asking Lego to end its partnership with Shell. In October 2014, Lego announced that it would not be renewing its promotional contract with Royal Dutch Shell. Greenpeace claimed the decision was in response to its campaigning.

Official Website

First launched in 1996, the Lego website provides many extra services beyond an online store and a product catalog. The website was and still is a social networking site that involves items, blueprints, ranks, and badges earned for completing certain tasks. There are also trophies called masterpieces, which allow users to progress to the next rank. The website has a built-in inbox that allows users to send prewritten messages to one another. By 2013, the Lego websites were attracting more than 20 million visitors a month.

Theme Parks and Discovery Centers

Merlin Entertainments operates eight Legoland amusement parks located in Denmark, England, Germany, California, Florida, Malaysia, United Arab Emirates, and Japan. A ninth was planned to open in 2021 in Goshen, New York, and a tenth in 2022 in China. Lego had limited experience in running theme parks. In 2005, the control of 70% of the Legoland parks was sold for $460 million to the Blackstone Group of New York, while the remaining 30% is still held by Lego Group.

There are also eight Legoland Discovery Centers: two in Germany, four in the United States, one in Japan, and one in the United Kingdom. Two Legoland Discovery Centers opened in 2013: one at the Westchester Ridge Hill shopping complex in Yonkers, New York, and one at the Vaughan Mills in Vaughan, Ontario, Canada. Another opened at the Meadowlands complex in East Rutherford, New Jersey, in 2014.

The target audience for the Legoland Discovery Center is families with young children, normally ages 3–12, though a typical location's average guest is about 7 years of age. Discovery Centers are located near other family-friendly attractions and dining establishments. In any given year, a single facility can host approximately 400,000 to 600,000 visitors.

A typical Legoland Discovery Center occupies approximately 30,000–35,000 ft^2 of floor area. Discovery Centers include models of local landmarks rendered in Lego bricks. Visitors can also learn how the Lego bricks are manufactured or partake in building classes taught by a Master Model Builder. Certain locations may also include movie theaters offering multiple showings throughout the day.

A few children's attractions, such as small rides and play fortresses, are also available. The centers can host birthday parties as well as scholastic and group functions and include restaurants and gift shops selling Lego merchandise.

Retail Stores

Lego decided to open its own stores and sell directly to consumers rather than have to rely upon the limited shelf space they were getting from retailers. Many of the retailers were not providing Lego with enough shelf space to display their branded products.

Lego operates 132 so-called "Lego Store" retail shops worldwide. The opening of each store is celebrated with weekend-long events in which a Master Model Builder creates, with the help of volunteers – most of whom are children – a larger-than-life Lego statue, which is then displayed at the new store for several weeks.

The stores are used to introduce entire families to the Lego experience. The stores interact with their customers to bring forth ideas for new Lego products.

Variations on Lego Themes

Since the 1950s, the Lego Group has released thousands of sets with a variety of themes, including space, robots, pirates, trains, Vikings, castles, dinosaurs, undersea exploration, and Wild West. Some of the classic themes that continue to the present day include Lego City (a line of sets depicting city life introduced in 1973) and Lego Technic (a line aimed at emulating complex machinery, introduced in 1977).

Over the years, Lego has licensed themes from numerous cartoon and film franchises and even some from video games. These include Batman, Indiana Jones, Pirates of the Caribbean, Harry Potter, Star Wars, and Minecraft. Although some of the licensed themes such as Lego Star Wars and Lego Indiana Jones had highly successful sales, Lego has expressed a desire to rely more

(Continued)

Case Study (Continued)

upon their own characters and classic themes, and less upon licensed themes related to movie releases. Lego created their own storylines and supporting characters that they believed would appeal to their audiences.

Minifigures

A Lego minifigure, commonly referred to as a minifig, is a small plastic figurine. They were first produced in 1978 and have been a great success, with more than four billion produced worldwide by 2006. Minifigures are usually found within Lego sets, although they are also sold separately as collectibles or custom-built in Lego stores. While some are named as specific characters, either licensed from film, television, and game franchises or of Lego's own creation, many are unnamed and are designed simply to fit within a certain theme (such as police officers, astronauts, or pirates). Minifigures are collected by both children and adults. They are highly customizable, and parts from different figures can be mixed and matched, resulting in many combinations.

For the 2012 Summer Olympics in London, Lego released a special Team GB Minifigures series exclusively in the United Kingdom to mark the opening of the games. For the 2016 Summer Olympics and 2016 Summer Paralympics in Rio de Janeiro, Lego released a kit with the Olympic and Paralympic mascots Vinicius and Tom.

One of the largest Lego sets commercially produced was a minifig-scaled edition of the Star Wars Millennium Falcon, which contained 5,195 pieces and was released in 2007. It was surpassed by a 5,922-piece Taj Mahal. A redesigned Millennium Falcon recently retook the top spot in 2017 with 7,541 pieces.

Robotic Themes

Lego also initiated a robotics line of toys called Mindstorms in 1999 and has continued to expand and update this range ever since. The roots of the product originate from a programable brick developed at the MIT Media Lab, and the name was taken from a paper by Seymour Papert, a computer scientist and educator who developed the educational theory of constructionism, and whose research was at times funded by the Lego Group.

The programable Lego brick, which is at the heart of these robotics' sets, has undergone several updates and redesigns. The set includes sensors that detect touch, light, sound, and ultrasonic waves.

The intelligent brick can be programmed using official software available for Windows and Mac computers and is downloaded onto the brick via Bluetooth or a USB cable. Several unofficial programs and compatible programming languages have been made to work with the brick, and many books have been written to support this community.

There are several robotics competitions that use the Lego robotics sets. They focus on middle school and high school competitions, like a Lego robotics tournament held at MIT. The tournaments focus on specific age groups, such as students aged 6–9 and 9–16. Students form teams and must use Lego-based robots to complete tasks. Students see this as a real-world engineering challenge. In 2010, there were 16,070 teams in more than 55 countries. The competition involved extensive use of Lego Mindstorms equipment, which was often pushed to its extreme limits. There is a strong community of professionals and hobbyists of all ages involved in the sharing of designs, programming techniques, creating third-party software and hardware, and contributing other ideas associated with Lego Mindstorms. Lego encourages sharing and peering by making software code available for downloading and by holding various contests and events. The overall benefit was that technology was bringing more adults to Lego products.

Integrated Experiences

The toys, minifigs, robotics, books, and accessories allowed customers to develop their own storylines when playing, including roleplaying, rather than just construction activities. This gave customers the opportunity to build a bridge between Lego's traditional toys and the digital world. If the minifigs and robotic themes were based upon TV shows and movies, customers could create their own storylines using their imagination and creativity.

Video Games

Lego has also branched out into the video game market with titles such as Lego Island, Lego Creator, and Lego Racers. Lego developed strategic partnerships to make games like Lego Star Wars, Lego Indiana Jones, Lego Batman, and many more, including the very well-received Lego Marvel Super Heroes game, featuring New York City as the overworld and including Marvel characters from the Avengers, the Fantastic Four, the X-Men, and more. More recently, Lego created a game based on The Lego Movie, due to its popularity. By 2013, more than 100 million copies of Lego video games were sold by their licensed partners.

(Continued)

Case Study (Continued)

Innovation Management: Plastic Construction Toy

From its inception through the late 1990s, Lego enjoyed a steady growth. In 1998, Lego suffered its first yearly loss and then hired a turnaround specialist as the CEO to get the company back on track. The company had been struggling with poor management, lack of a strategic focus, and a disconnection from their customers when Lego was not responsive to customers' needs. There were innovations and some successes, but the innovation management process appeared unstructured. Unable to turn the company around, a new CEO was hired in January 2004. Lego suffered another significant loss in 2004 and was on the verge of bankruptcy.

The marketplace had changed. Children were growing older at a faster rate. Lego's target market of boys aged four to nine was turning to video games and web-based activities. Other toy manufacturers were working with licensing partners and were becoming serious threats to capture Lego's core customers.

Turning the company around was difficult. Lego had allowed innovation to be uncontrolled and did not know whether the focus should be on sustaining innovation mainly around the Lego bricks, disruptive innovation for new products, or both. It is questionable whether the innovations were linked to strategic business objectives. The number of new products had increased from 6,000 to 14,200. Lego failed to realize that too much innovation can be unhealthy. As stated by management expert Peter Drucker, "There is nothing so useless as doing efficiently that which should not be done at all."

Lego was struggling with costs. They did not know the costs of many individual sets and had trouble identifying which products and product lines were profitable.[6]

Lego appeared to have suffered from some of the traits common to many privately held companies regarding innovation. The starting point was with Lego's business model. It is not uncommon for privately held companies to focus on new product development without recognizing that business model innovation may also be necessary. Without having shareholders, the company can lack a financial and operative control system to the point where cost management can get out of control. All ideas for new products may be internally generated, based upon the whims of management, and with no involvement by customers.

---

6 Adapted from Ville Kilkku, "Sustaining innovation and disruptive innovation – case Lego," June 2014. http://www.kilkku.com/blog/2014/06/sustaining-innovation-and-disruptive-innovation-case-lego/. Accessed February 2020.

The company may rush into the launch of new products without proper prototype development and testing. Executives may fall in love with the existing products and refuse to see the benefits of licensing their intellectual property. Executives may not see the benefits of lowering costs by outsourcing production to lower-cost organizations. If Lego were to survive, the effectiveness of its product development process would need to change. Growing the business is the right thing to do, but it must be accompanied by business model transformation if necessary. Only by radically reimagining and speeding up the process could Lego create breakthrough new toy ideas and save the company.[7] There were, of course, other things Lego had to do as well.

The turning point occurred with the launch of Mindstorms. Within three weeks after the launch, more than 1000 advanced users – in a campaign coordinated on the web – had hacked into the software that came with the construction toys to make unauthorized modifications with new functions.[8] The hackers had actually improved the product to the point where more units were being sold. Lego's original thoughts were that the hacking was illegal and done without permission. But Lego soon realized that the product was attractive to customers over the age of 18. Lego was on the verge of finding a new customer base and decided that it would be best not to fight with hackers but instead to harness their knowledge and creativity to improve the products.

Lego quickly realized the benefits of open innovation. Lego could tap into the brainpower and imagination of others rather than relying solely upon its own R&D group. The possibilities were endless. Lego was now about to reverse its downward trajectory and return to profitability. Lego's culture and business model were changing. Lego was now listening to its customers. Initially, management feared that this would slow down the product development processes but soon realized that their fears were unwarranted.

Lego's turnaround strategy came from engaging its expansive customer base. The goal was to generate customer feedback on a small scale before making substantial investments, illustrating Lego's philosophy that "people don't have to work for us to work with us."[9] To further this practice, the company launched Lego Ideas, an online crowdsourcing platform, allowing

(Continued)

7 Jonathon Ringen, "How Lego Became the Apple of Toys," Fast Company, June 15, 2015.
8 Adapted from "Lego Success Built on Open Innovation," November 13, 2018. https://www.idea-connection.com/open-innovation-success/Lego-Success-Built-on-Open-Innovation-00258.html. Accessed February 2020.
9 Leadership Network, "5 Sustainable Innovation Practices that Saved Lego," Innovation Management, November 7, 2016.

Case Study (Continued)

customers to share and to vote for ideas they wish to see as additions to the product line. Lego Ideas yielded hundreds of suggestions annually, employing social media to generate actionable data. Focusing on products that would sell, Lego was able to reach new audiences through its extensive physical footprint and brand awareness.[10]

Lego introduced several programs to make it easy for consumers to work with the company. The company launched the Ambassador Program that provided a direct way for it to access new ideas from its community.[11] A new platform named Lego Cuusoo was launched to allow fans to upload designs. If a design received 10,000 votes from the community members, Lego agreed to consider it for production. This process maximized the possibility that a product would have mass appeal.[12] Another open-source platform was Adult Fans of Lego (AFOL). The consumers could work as lead developers with Lego personnel.

Lego also created a Future Lab to control internal and customer-generated innovation ideas at Lego. This innovation lab was tasked with inventing the future of play, a large part of which was identifying growth opportunities and ensuring that Lego stays ahead of the curve. They strived to do things not otherwise done and to introduce radical innovation without jeopardizing the core business and value propositions of the Lego brand.[13] Lego now had design talent spread across the world and all of these new programs created by Lego were focused on connecting the ideas with the product development teams. Lego had changed their business model to become closer to their customers.

Innovation Life Cycle Phases

Primary concept and development work occurs at the Billund headquarters, where the company employs product designers. The company also has smaller

10 Adapted from Jaclyn Markowitz, "Open Innovation at Lego: The Back Beat in 'Everything is Awesome,'" November 13, 2018. https://digital.hbs.edu/platform-rctom/submission/open-innovation-at-lego-the-back-beat-in-everything-is-awesome/. Accessed February 2020

11 Adapted from "Bricks & Code: Open Innovation at Lego Group," November 13, 2018. https://digital.hbs.edu/platform-rctom/submission/bricks-code-open-innovation-at-lego-group/. Accessed February 2020.

12 Yun Mi Antorini, Albert M. Muniz, Jr., and Tormod Askildsen, "Collaborating with Customer Communities," MIT Sloan Management Review, Spring 2012.

13 Adapted from Michael Fearne, "Lego Future Lab: The Rebels of Innovation at Lego," 2019. https:// michaelfearne.com/lego-future-lab-the-rebels-of-innovation-at-lego/. Accessed March 2020. The website contains several good blogs on innovation practices at Lego.

design offices in the UK, Spain, Germany, and Japan that are tasked with developing products aimed specifically at these markets. Even though Lego has offices dispersed around the world, there is still commonality and interconnectivity among their products whereby all innovation projects appear to contain the same or similar life cycle phases even though some projects have different phases.

The average development period for a new product is around 12 months, split into three life cycle phases. The first phase is to identify market trends and developments, including contact by the designers directly with the market. Some designers are stationed in toy shops, especially close to holidays, while others interview children and their parents. The second phase is the design and development of the product based upon the results of the first phase.

The design teams use 3D modeling software to generate CAD drawings from initial design sketches. The designs are then prototyped using an in-house stereolithography machine. These prototypes are presented to the entire project team for comment and testing by parents and children during the "validation" process. Designs may then be altered in accordance with the results from the focus groups. Virtual models of completed Lego products are built concurrently with the writing of the user instructions. Completed CAD models are also used in the wider organization, for marketing and packaging.

The third life cycle phase is the actual commercialization of the product. After product launch, Lego interacts closely with the consumers for improvements to the products using incremental innovation as well as seeking out ideas for other similar and nonsimilar products that would require radical innovation.

Creativity and brainstorming are critical innovation skills in all life cycle phases at Lego. The brainstorming and creativity extend to their user base as well. In May 2011, Space Shuttle Endeavour mission STS-134 brought 13 Lego kits to the International Space Station, where astronauts, built models to see how they would react in microgravity, as a part of the Lego Bricks in Space program. In May 2013, the largest model ever created was displayed in New York City and was made of over 5 million bricks; a 1 : 1 scale model of an X-wing fighter. Other records include a 112-ft tower and a 2.5-mile railway.

Innovation Management Lessons Learned

One of the risks with family-owned businesses when things are going well financially is that they tend to become complacent, with an attitude of

(Continued)

Case Study (Continued)

"Let's leave well enough alone" or "The same old way will work for years to come." The only innovation that is considered is then incremental innovation, and lessons learned may not be shared across the entire company. Lego did not fall into this trap. Some of the things that Lego appeared to understand were:[14]

- Lego must foster an innovation culture.
- Innovations may require changes to the firm's business model.
- Survivability is based upon using multiple forms of innovation, even though emphasis is placed upon incremental and radical innovation.
- Different levels of innovation by different groups must be allowed to improve the product success rate.
- Radical, incremental, and other forms of innovation may follow different life cycle phases.
- Radical innovation is difficult and control systems, gates, and checkpoints are necessary.
- Innovation processes must be decentralized and innovation teams must have some freedom in selecting the best approach, such as deciding between using a waterfall or agile project management approach.
- Innovation teams must have the choice of tools to be used on their projects.
- All innovation projects must be based upon the firm's core values.
- Not all Lego's products will be successful.
- Lego's future must include crowdsourcing practices and maintaining a constant dialogue with its customers.
- Lego must get close to the customers it serves, not just for idea generation, but to ensure that a market exists for the outcome of projects.
- Lego must understand the changes that are taking place in the needs and behaviors of its customers.
- Lego must realize that they have multiple customer bases, especially among older users.
- Many of Lego's customers want to participate as cocreators, if just in providing ideas or in actual participation in product development.

14 Some of the lessons learned have been adapted from Divina Paredes and David Gram, "Lego: An insider's guide to radical innovation," CIO, June 6, 2017. For a more detailed description of the innovation processes at Lego, see Robertson, David. Brick by Brick: How LEGO Rewrote the Rules of Innovation and Conquered the Global Toy Industry, New York: Random House, 2013.

- To maximize ideas, Lego must provide a mechanism whereby customers acting as cocreators can communicate and exchange ideas with one another.
- New product testing, pilot studies, prototyping, and experimentation will be necessities.
- Design thinking must be part of innovation processes.

Epilogue

The Lego case is an example of the complexities of managing innovation to sustain a global brand. There are numerous challenges, including expanding the brand into new ventures such as games, videos, movies, apparel and accessories, company-owned stores, and licensing agreements. Should the company focus more on incremental or radical innovations? For innovation to be successful, should the company centralize or decentralize operations? Should the entire company realize the strategic vision? How will each innovation impact the firm's business model? These issues must be addressed continuously.

The Lego Group announced on September 4, 2017, its intention to cut 1,400 jobs following reduced revenue and profit in the first half of the year, the first reported decrease in 13 years. The revenue losses appear to be the result of a more competitive environment, where the company has to compete not only against its traditional rivals such as Mattel and Hasbro, but also against technology companies such as Sony and Microsoft as more children are using mobile devices for entertainment. However, some insiders at the Lego Group believe that Lego has become complacent due to recent yearly earnings, may have lost its entrepreneurial and innovative spirit, and that it may take a few years to recover. History has shown us that Lego has the capability to overcome these hurdles.

Questions

1. What are some of the innovation project management critical issues that may (or may not) be unique to privately held companies, as opposed to publicly held firms?

2. Should project managers participate in market research studies to determine who the customers are?

3. Should project managers participate in follow-up market research studies to determine how well the customers like the products?

(Continued)

Case Study (Continued)

4. Do companies have core values and, if so, what might cause them to change, as in the Lego case?
5. Should innovation project managers understand licensing agreements?
6. Should project managers be concerned about trademarks, intellectual property, and environmental issues? If so, what depth of knowledge should exist?
7. Can websites be of benefit to innovation project managers?
8. Were the life cycle phases used in the Lego case traditional life cycle phases for innovation projects?
9. What types of innovation were used at Lego?
10. Do brand management activities place limitations on whether innovation practices are centralized or decentralized?

Reflections: This Lego case study is a multifaceted example of what it takes to lead a global experience-driven culture. The common thread to the journey of Lego has been its culture and set of core values. The learning aspect of being experience driven was shining throughout the flow of the case study. As an organization, like Lego, goes through its cycles of connecting with the market and clients, it learns a great number of valuable lessons. Not only are the lessons focused on innovation and how to address the needs of different innovation life cycles and control mechanisms, depending on products and markets, but also lessons enable the understanding of how to adjust the business model, and fast, to sustain growth and tilt as needed.

There are many interesting moments in time when unexpected events would contribute to wonderful growth opportunities. In Lego's example, changing its attitude toward hackers and taking that as opportunity to create a much more open culture, could be considered another great shift milestone. Having an open ecosystem is critical for effective experiencing. Having multiple strategies, products, and platforms to connect with the customer's voice directly, is a key differentiator for the future organizational cultures.

> **Tip**
> For designing and building experience-driven features, organizations must be open to continuously learn what matters most to their customers and markets.

5.2 Continual Adapting

The Lego journey provides a very diverse list of business and management lessons. Most of all, it highlights the importance of proper governance, need for inspiring leadership, and an unlimited appetite for learning. Ensuring that the organization is humble enough to learn from markets, customers, and competitors, is a fundamental key growth attribute.

Especially when an organization is dealing with answering key questions about where the growth is in relation to the target client, coupled with trying to figure out what and how much innovation is right, adapting is needed. This requires an effective model of sensing and a strong amount of collaboration across stakeholders and regions. Figure 5.2 highlights some of the key building blocks for this effective experience, while addressing what features are fit for an organization and its growth needs and confirms the importance of continual adapting capability.

The first building block in continual adapting is a critical one. Cultures will succeed in driving experimentation when they enable adjusting their organizational strategic focus continually as needed. The classic statement "success could be our worst enemy," could apply here. Many instances over the journey of an organization, when things are going really well, this success could result in progressing without true links to a holistic long-range strategy, thus likely leading to situations when the future of the organization could be in jeopardy.

As highlighted in Figure 5.3 by the PwC annual CEO survey, a high percentage of CEOs are concerned about their organizations' ability to sustain their growth into the future and whether they last past a certain period of time. This shows red

Figure 5.2 Continual Adapting Building Blocks.

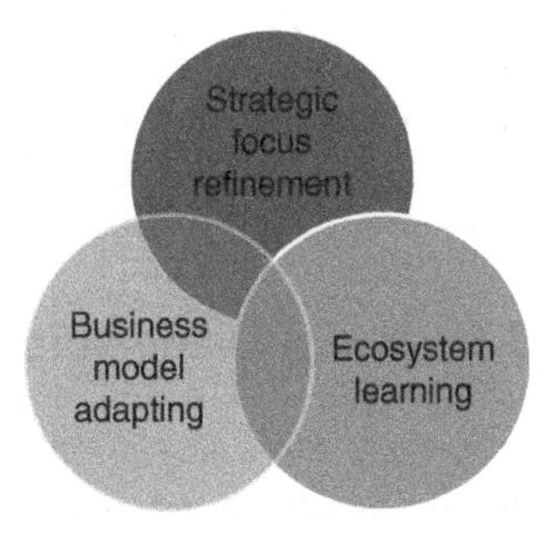

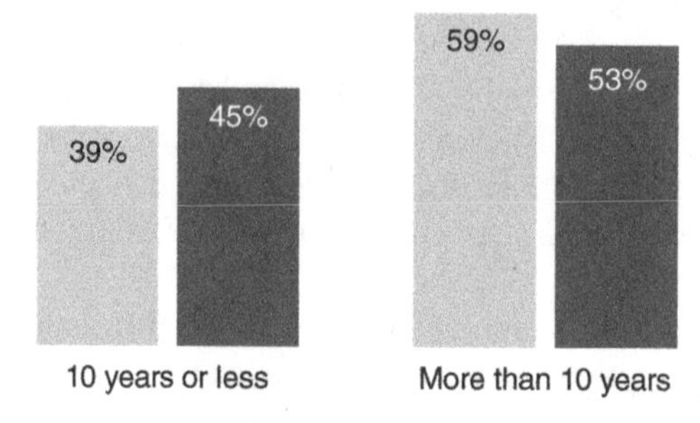

Figure 5.3 The Need for Adapting.

flags potentially pertaining to the right amount and type of innovation, adaptability in business model adjustments, and whether true experiencing remains to be the norm of doing business.

The second building block of continual ecosystem learning is a strong differentiator for organizations and teams that exhibit strength in continual adapting. The growth mindset and the humility that are required to learn from every stakeholder and partner in the ecosystem, are wonderful attributes to remaining relevant and impactful.

The 3rd building block of business model adaptation becomes a natural outcome of effective strategy refinement and the openness for learning. To succeed in establishing and benefiting from this building block, a high transparency in the culture needs to prevail in order to ensure that we don't miss the signs, or think that it is not about how we do business and just about how much we create and innovate.

> **Tip**
> Continual adaptation increases the chances that we achieve effective experiences across our organizational and teams' cultures.

5.3 Social Leadership

In our book, Kerzner et al. (2022), we explore the many dimensions of social leadership and how this is becoming a model for leading into the future. For this growing focus on experiencing, and as highlighted by the Lego case study, this model of leading creates a nurturing environment for experiencing to flourish.

Leaders generally face a dilemma of whether to be an authoritarian (i.e., directive) or social leader. Authoritarian leaders expect team members to comply with instructions and may provide team members limited opportunities to be creative and anticipate problems. Social leadership focuses on collaboration, trust, and empowerment. Changing leadership styles during a project from social to authoritarian can create confusion and alienate team members when they believe they have lost their empowerment and no longer trust the project manager. Going from authoritarian to social can be equally as bad if team members believe that this is temporary and can change back quickly. Team members may have to live with fear or uncertainty, which generates mistrust and may encourage team members to resist changes. The result can be a negative impact on project outcomes.

The good news is that more studies are being conducted that focus on identifying effective leadership traits (Muller and Turner 19, 20) and Shao and Muller (26). These studies identified leadership competencies categorized into emotional competencies, managerial competencies, and intellectual competencies. These types of studies open the door for a better understanding of effective social project management leadership and may eliminate or explain the inconsistent findings in earlier research.

Projects are managed by people, not tools. There must be a concerted effort by companies, as well as the project management community at large, to recognize the growing importance and need for effective social project management leadership supported by an understanding of servant leadership concepts and emotional labor.

With the growth in project management metrics and the ability to measure anything and everything, we believe today that we can establish tangible and/or intangible metrics that can measure the success of social project management leadership. The capturing of best practices and lessons learned in the future will also include the effectiveness of project management leadership and the impact on project teams.

The application of social leadership in organizational work builds the right amount of motivation across initiatives and teams focused on innovating the future direction of the organization. Across the many changes over the life of Lego, it was obvious the need for the organization to reinvent how it leads, connects across markets, adjusts to technological shifts, and most importantly operates from a humble leadership position that is hungry to learn and adjust to grow and create sustainable impact.

> **Tip**
> Social leadership offers an opportunity to strengthen the organization's ability to look at its portfolio of innovations differently and to adapt faster.

5.4 Creating a New Mindset

The need for effective experiencing could be blocked if a new mindset does not prevail. Figure 5.4 highlights some of these obstacles that could prevent the birth of the new mindset required for the experience-driven culture.

Signs of some of these obstacles along the way of organizations' journeys, could affect the necessary experience. An organization could struggle with changes that are not part of its history, original purpose, or how it got here. To change mindset, this obstacle could stop the organization in its tracks. The key to effective change is to exhibit flexibility to other ideas beyond the natural comfort zone of the leadership and the teams.

When a portfolio management process is focused on generating as many products and solutions, regardless of the linkages to the strategic impact or the change that is deemed necessary on the market or the customers, this would show a weakness that would be hard to overcome. It would require a fresh view of how strategic choices are being made and the right data behind some of the mindset shifts that are needed to support a more effective decision-making process.

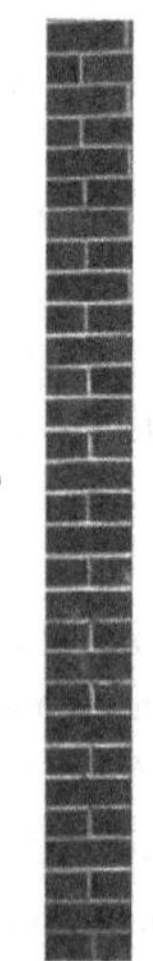

Figure 5.4 Mindset Obstacles to Overcome.

The 3rd obstacle of "protecting the brand at all costs" is what could lead to a closed organization, build unrealistic level of fear of competitors and partners, and worst of all minimize the chances to create an open ecosystem. Of course, protecting the brand is a key responsibility, yet the actions and behaviors need to ensure looking at this holistically and taking on the necessary risks necessary to change the mindset and support sustained growth.

The last obstacle in the figure is potentially one of the worst ones, as it could directly lead to losing relevancy. When the ego gets in the way, or the success view assumes that no change or learning is needed, the organization stops to be relevant.

It might take few more months or years to show, yet ultimately something will be lost in the secret sauce necessary for the brand to have the effect that it used to have, or worse that it could potentially have had, yet will not be realized because the teams have not been experiencing enough.

Tip

For effective experience to succeed, a new mindset, matters. Building this mindset requires risk-taking and stepping into the realm of the possible with learning at the core.

5.5 Sustaining Creativity

One of the most critical aspects of maintaining effective experience is to sustain creativity as a dominant quality for the organization and across its teams. This assumes that many of the healthy fundamentals are in place to support openness, critical conversations, deep learning, access to powerful data, and the balance addressed earlier in this work between humans and digitalization.

As in Figure 5.5, the need to have an attentive mind, especially in this era of extensive distractions, is very important for creativity. At the center of this attentive mind, is the willingness to serve. As in the case of servant leadership, or as highlighted earlier in this work around social leadership, it is key to have this attitude to ignite creativity. When this is coupled with the love for the mission, and the anticipated impact that the outcomes of the work is expected to create, creativity has a high potential of steadily flowing and of being sustained.

It is easy to have individual instances when some level of creativity is shown, yet the creation of sustained creativity, and the protection of the environment that supports that, could be a challenge. It is a critical shift in how the way of working and experiencing across the culture prevails. Elements of caring and encouragement are equally valuable ingredients for this sustainability to remain.

Figure 5.5 The Attentive Mind. Credit: johnhain/Pixabay.

When creativity is sustained, the potential for effective experiencing will increase. The track record for creativity's impact could be shown in the success stories that would be replicated. As in how excellence and maturity are achieved, a set of successful patterns for thinking start becoming more consistent and this creates a contagious ripple effect across teams.

The future cultures will benefit from attentive minds across the organizational teams. Sustaining creativity requires relentless commitment of the management team and needs to be supported by an encouraging environment, where listening and respect dominate. Replicating this behavior is critically important, and ideally, this should also be linked to the right mix of metrics, used to gauge the success of the organization, and its strategic choices.

Tip

Responsible leadership should ensure ways to sustain creativity. This is a core ingredient for experiencing and truly ignites the organization's energy for innovation.

Reference

Kerzner, H., Zeitoun, A., and Vargas, R. (2022). *Project Management Next Generation: The Pillars for Organizational Excellence*, 1e. Wiley.

Review Questions

Parentheses () are used for Multiple Choice, when one answer is correct. Brackets [] are used for Multiple Answer, when many answers are correct.

1 Which of the following is an effective enabler for designing experience-driven features?

( ) Increased focus on actions.
( ) Successful brainstorming.
( ) Full reliance on management's direction.
( ) Moving to solutions fast.

2 What are good signs of effective experiencing? Choose all that apply.

[ ] Ability to rethink the mix of innovations.
[ ] Being determined to keep to tradition.
[ ] Multiple ways to listen to customers.
[ ] Ability to adapt fast.

3 What is the greatest value of having multiple ways to capture customer voices?

( ) Check the box on a business requirement.

(Continued)

(Continued)

( ) Depends mainly on the customer to design the operating model.
( ) Making sure products will have a market relevance.
( ) Enabling teams to confirm their views.

4 What is fundamental way to maintain the organizational consistency?
( ) Put many processes in place.
( ) It is not necessary to maintain consistency.
( ) Depends on the organizational core values.
( ) Adjusting with any new strategic insights.

5 What are useful building blocks for continual adapting? Choose all that apply.
[ ] Refining strategic focus.
[ ] Project management processes.
[ ] Ecosystem learning.
[ ] Focus on utilizing one business model.

6 What is a key attribute to social leadership?
( ) Keep an authoritarian role over the project team.
( ) Focus on collaboration, trust, and empowerment.
( ) Increase uncertainty across project teams.
( ) Spend more time generating progress reports.

7 What is an important quality to sustain creativity in how we work?
( ) Encourage more multitasking.
( ) Having a prevailing service mindset.
( ) Keep celebrating till the end of the projects.
( ) Ensure making decisions fast.

6

Human Connection

The future is **human-to-human**. Human connection is the differentiator in the digital age. Excellence in leading across cultures will require stability at the helm that provides trust and confidence across teams and stakeholders. Adopting technology should be done responsibly to secure higher efficiencies and allowing the project and program leaders and their teams to produce higher potential value from their portfolio of work.

This human-to-human connection is at the core of the responsible cultures that naturally commit to a growth mindset and learning. Experience-driven culture is committed to balance, continual capabilities evolvement, and a strategic way of working that is built around stronger strategic insights.

Evolving the new human will come from new ways of thinking, working, and collaborating. Teams will be supported on the path of operational excellence and classical models of leading and working will be further disrupted. Inspiring leaders will be providing a steady hand in changing and transforming the organizations of the future while elevating our world.

Key Learnings

- Understand the power of bridging and the critical value it enables in crossing boundaries and aligning focus.
- Get intentional about how to design communications for impact and to remain creative and innovative across the experiencing culture.
- Learn key principles for how to use portfolio principles to sustain excellence in delivering outcomes across organizations and teams.
- Learn the art of mastering how being experience-driven opens the door for generating effective ideas, engaging, and the achievement of sustained outcomes.
- Develop your capabilities to be the inspiring leader who releases the energy for teams and individuals toward transformational innovation and projects' value achievement.

Creating Experience-Driven Organizational Culture: How to Drive Transformative Change with Project and Portfolio Management, First Edition. Al Zeitoun.

6.1 The Power of Bridging

The power of bridging is a foundational way for effective human connection. One could view this from a context of communicating and finding ways to get one's ideas across and potentially win an argument or negotiate a reasonable outcome. Yet the implications of this power are vaster. In the experience-driven culture focus of this work, this power opens the doors to strong collaboration. For the classical challenge of projects and programs stemming from silos across the organization, this power is core to the horizontal ways of working needed for initiatives' success.

With the high focus on innovation, as highlighted in the previous chapter, bridging allows a fluid exchange of ideas along the lines of the sustained creativity previously discussed. Breakthrough innovations could materialize more consistently if organizations could strengthen their human bridging power. When teams build a strong foundation of trust and empower their bridging with the right data, they become competitive and able to tackle complex challenges easier and faster.

The bridging is also core to problem-solving when teams are able to use visuals to connect across their ideas. Whether we use a 2 × 2 Matrix to illustrate Strengths, Weaknesses, Opportunities, and Threats (SWOT), or a concept board for the team to establish the proper mapping across concepts and views, all of this energizes the human connection and possibly closes gaps that might have remained an execution challenge.

Figure 6.1 is a simple illustration that shows how teams and teamwork are at the core of the bridging connection. To achieve the full power of bridging, the leader should create the right supporting environment where all voices are respected and heard. The inclusion of ideas that is needed here, helps in establishing the learning principles that are needed for the experience-driven cultures that we esteem. As in the figure, it takes persistence and patience, connecting one building block at a time to find the right and most suitable way to bridge across the diverse ideas and sometimes opposing views that dominate groups of stakeholders in the work environment and across the ecosystem.

> **Tip**
> In building connected organizational cultures, the power of bridging is a superpower for crossing human connection barriers.

The following case study highlights the missing elements of this crucial bridging capability. It might be a good example of what not to do if your goal is to create a bridge to your key project stakeholders. Bridging requires listening and a two-way dialogue, in a safe and respectful environment.

Figure 6.1 Bridging Connection.

Case Study

Kemko Manufacturing[1]

Background

Kemko Manufacturing was a 50-year-old company that had a reputation for manufacturing high-quality household appliances. Kemko's growth was rapid during the 1990s. It grew by acquiring other companies. Kemko now had more than 25 manufacturing plants throughout the United States, Europe, and Asia.

Originally, each manufacturing plant that was acquired wanted to maintain its own culture, and quite often each was allowed to remain autonomous from corporate at Kemko provided that work was progressing as planned. But as Kemko began acquiring more companies, growing pains made it almost impossible to allow each plant to remain autonomous.

Each company had its own way of handling raw material procurement and inventory control. All purchase requests above a certain dollar value had to

(Continued)

1 Kerzner, 2022/John Wiley & Sons.

Case Study (Continued)

be approved by corporate. At corporate, there was often confusion over the information in all of the forms since each plant had its own documentation for procurement. Corporate was afraid that, unless it established a standardized procurement and inventory control system across all plants, cash flow problems and loss of corporate control over inventory could take its toll in the near future.

Project is Initiated

Because of the importance of the project, senior management asked Janet Adams, director of information technology (IT), to take control of the project personally. Janet had more than 30 years of experience in IT and fully understood how scope creep can create havoc on a large project.

Janet selected her team from IT and set up an initial kickoff date for the project. In addition to the mandatory presence of all of her team members, she also demanded that each manufacturing plant assign at least one representative and that all plant representatives be in attendance at the kickoff meeting. At the meeting, Janet said:

I asked all of you here because I want you to have a clear understanding of how I intend to manage this project. Our executives have given us a timetable for this project and my greatest fear is scope creep. "Scope creep" is the growth of or enhancements to the project's scope as the project is being developed. On many of our other projects, scope creep has lengthened the project and driven up the cost. I know that scope creep isn't always evil and that it can happen in any life cycle phase.

The reason why I have asked all of the plant representatives to attend this meeting is because of the dangers of scope creep. Scope creep has many causes, but it is generally the failure of effective up-front planning. When scope creep exists, people generally argue that it is a natural occurrence and we must accept the fact that it will happen. That's unacceptable to me!

There will be no scope changes on this project, and I really mean it when I say this. The plant representatives must meet on their own and provide us with a detailed requirements package. I will not allow the project to officially begin until we have a detailed listing of the requirements. My team will provide you with some guidance, as needed, in preparing the requirements.

No scope changes will be allowed once the project begins. I know that there may be some requests for scope changes, but all requests will be bundled together and worked on later as an enhancement project. This project will be implemented according to the original set of requirements. If I were to allow scope changes to occur, this project would run forever. I know some of you do not like this, but this is the way it will be on this project.

There was dead silence in the room. Janet could tell from the expressions on the faces of the plant representatives that they were displeased with her comments. Some of the plants were under the impression that the IT group was supposed to prepare the requirements package. Now Janet had transferred the responsibility to them, the user group, and they were not happy. Janet made it clear that user involvement would be essential for the preparation of the requirements.

After a few minutes of silence, the plant representatives said that they were willing to do this and it would be done correctly. Many of the representatives understood user requirements documentation. They would work together and come to an agreement on the requirements. Janet again stated that her team would support the plant representatives but that the burden of responsibility would rest solely on the plants. The plants would get what they ask for and nothing more. Therefore, they must be quite clear up front in their requirements.

While Janet was lecturing to the plant representatives, the IT team members were just sitting back smiling. Their job was about to become easier, or at least they thought so. Janet then addressed the IT team members:

Now I want to address the IT personnel. The reason why we are all in attendance at this meeting is because I want the plant representatives to hear what I have to say to the IT team. In the past, the IT teams have not been without some blame for scope creep and schedule elongation. So, here are my comments for the IT personnel:

- It is the IT team's responsibility to make sure that they understand the requirements as prepared by the plant representatives. Do not come back to me later telling me that you did not understand the requirements because they were poorly defined. I am going to ask every IT team member to sign a document stating that they have read over the requirements and fully understand them.
- Perfectionism is not necessary. All I want you to do is to get the job done.
- In the past we have been plagued with "featuritis," where many of you have added in your own bells and whistles unnecessarily. If that happens on this project, I will personally view this as a failure by you, and it will reflect in your next performance review.
- Sometimes people believe that a project like this will advance their career, especially if they look for perfectionism and bells and whistles. Trust me when I tell you this can have the opposite effect.

(Continued)

Case Study (Continued)

- Backdoor politics will not be allowed. If any of the plant representatives come to you looking for ways to sneak in scope changes, I want to know about it. And if you make the changes without my permission, you may not be working for me much longer.
- I, and only I, have signature authority for scope changes.
- This project will be executed using detailed planning rather than rolling wave or progressive planning. We should be able to do this once we have clearly defined requirements.

Now, are there any questions from anyone?

The battle lines were now drawn. Some believed that it was Janet against the team, but most understood her need to do this. However, whether the project could work this way was still questionable.

Questions

1. Was Janet correct in the comments she made to the plant representatives?
2. Was Janet correct in the comments she made to the IT team members?
3. Is it always better on IT projects to make changes using enhancement projects or should we allow changes to be made as we go along?
4. What is your best guess on what happened?

Reflections: The case study showed that Janet had one major priority, namely controlling scope creep. The big challenge with this is that the team could win this battle and lose the war. Namely, this means that a joint view of what success looks like, when missing, could have more drastic effects than just scope creep. An authoritarian style of leadership dominated the case and any interest in the business or the IT team's opinions was lacking.

Also, the ways of working, and the likelihood of them including the experience-driven principles, was lacking. It looked like a pure classical waterfall approach was her goal and that any iterative work will not be allowed. In tomorrow's organizations, it is going to be rare that such an extreme approach to executing work will be reliable in dealing with the increasing uncertainties, coupled with greater market demands and expanded regulations.

> **Tip**
> Approaching high visibility programs requires management to be selective in their choices of the leader and the need for possessing bridging abilities.

6.2 Communicating with Impact

The movie, **"We are Marshall"** reflects the story of the small town of Huntington, West Virginia that suffered the loss of its entire school football team, staff members, and boosters who were in a plane that crashed. Following this tragedy, Coach Jack was hired as the new head coach.

The story shows his international communication style that he needed to utilize to rebuild Marshall's Thundering Herd and the entire community. He had to exercise the highest level of empathy to heal the grieving community while delivering the results of the football team. Jack used the power of bridging, opened the line of collaboration and dialogue, and managed to have the town regain the trust in itself and its team. Impactful communication could lead to great outcomes.

Just like in this movie, transformational projects rely heavily on the strength of communication. With the anticipated changes ahead of the project team in these projects, and with the potential of complexity across stakeholders and technology, clarity and simplicity of communications are critical.

Across a portfolio of projects, there is also the likelihood that a bubble could exist across the teams. This means that silos in understanding get built. When clarity of language used to connect across the teams is lacking, communication fails. Bubbles have to be clarified or better destroyed for the intentionality of communication to get across the bridge.

This requires that we take communications to the lowest level of complexity. The assurance that every key stakeholder is on the same page is part of this intentionality. Using the right and well-understood vocabulary is a must, and leaders in the future need to be sensitive to clear language they use and become more empathetic to the various needs across the possible silos.

Figure 6.2 highlights the delicate balance in the new mix of communications. With digitalization, data plays an instrumental role in successful connecting. This adds to the responsibility of the leader in being intentional in the use of data to tell the compelling stories required to create the needed change as an outcome of that communication. This starts with ensuring data accuracy, comprehending the context behind the data, confirming the proper meanings we extract from it, and the potential trends or assumptions we could build from it. Additionally, leaders who want to become more intentional in how they connect must take time to design the proper flow and rhythm of the communication necessary to drive the anticipated outcomes. This is another strong example of the mix between science and art.

Figure 6.2 Impactful Communication. Credit: TungArt7/Pixabay.

Tip
Intentional communication is an orchestration of a design that creates the required change outcomes. Leaders will thrive in the future when data supports impact creation

6.3 The Portfolio Success Link

Successfully executing a portfolio of projects is a sweet spot where the capacities and capabilities meet. Organizationally, the management team should have a clear view of the capacity of the organization to handle adding potential investments/initiatives to the mix. On getting this accomplished side, this is where the capabilities of people and technology come together to address those prioritized initiatives. A classic supply and demand usually is in the balance leading to overutilization and underutilization scenarios.

This is where experience-driven cultures that are powered by artificial intelligence (AI) help us predict these gaps, better plan for them, and achieve the most homogenous utilization of the organizational capacity. The portfolio success link could then refer to how well an organization makes strategic choices that cascade clearly across the organization without gaps in communications.

Ideally, this cascade has already been tested in the virtual world, so that the risks associated with different approaches, customization of the ways of working, balancing cultures and geographies involved, and pulling together the ideal mix of skills, are all getting a higher likelihood of success. A Portfolio Management Office could play a meaningful role in the proper orchestration of these changing portfolio dynamics.

in our article, Kerzner and Zeitoun (2023), we highlighted some of the key ingredients necessary to crack the excellence code and sustain the excellence path.

Historically, most project management textbooks defined excellence as a continuous stream of successfully managed projects. Many of the definitions of excellence focused more so on how the contractors used the tools and techniques of project and program management rather than the impact on the customers and stakeholders. Articles have been written discussing how work breakdown structures, statements of work, and capturing best practices lead to excellence.

Today we realize that there are many components of excellence. Excellence is now being defined in business as well as behavioral terms. Behavioral excellence has become a much more critical component than in the past because of AI. Excellence is not just in implementing the tools and techniques effectively, but also in understanding the true value of project management as seen by customers and stakeholders.

Excellence organizationally and personally is now a journey. In the context of project management, going on that journey has been changing over time and especially fast in the last few years. The mega technology disruptions have contributed to the change and the recent potentially positive scaling with generative AI is no exception.

In the article, we tackled a few elements pertaining to the changing views and approaches to excellence in delivering the promises of projects and programs and questioned some of the traditional views of leading and driving teamwork in this digital age. We would like to crack the code on how excellence has changed over time and what future leaders would need to equip themselves with as part of their newly expanded toolbox.

Excellence in project delivery has usually been tied to consistency in utilization of certain practices and supporting behaviors time and again. This is not enough in the future as the patterns and the ways of working will continue to change at a fast pace and what might be a best practice today might be challenged as the project team goes to work in the morning. This intense level of adaptability is empowered by technology and the merging of the virtual and real worlds has become the norm. Technology is changing not only the content of the project management forms, guidelines, templates, and checklists we are using today but also how they will be used in the future. Simply stated, technology is helping us understand the true value of excellence in project management.

Tomorrow's portfolio links are highly technology-dependent. The more power from data we utilize, the easier the balance of capacity and capability will be achieved thus minimizing allocation gaps and mistakes. This consistency empowers leaders. This simplifies the working environment and ways of working selection for the teams. The experience that happens with data makes these future cultures more efficient and effective in realizing initiatives' value.

Tip

Portfolio success link hinges on the right balance between capacity and capability. Stroger link means enhanced choice-making that cascades clearly across the organization.

6.4 Experience-Driven Mastery

Achieving mystery in being experience-driven is an ideal state. As covered previously in this work, it is more of a journey toward excellence, where the goal becomes the creation of high level of consistency in how effective experience is created and supported. The strength of this muscle centers around the ability to adapt and handle change. In many cases, it is actually a soaring appetite and interest in change given the potential upside that change could bring. This is similar to the "State of the Culture" commitment to continually revisit the kind of culture the organization builds, the associated success metrics, and the assurance of the ongoing practices toward that mastery muscle.

In one of the eight hypotheses behind this work, the following hypothesis was formulated: ***Change management is increasingly a vital business muscle.***

The following question was asked: *In your experience, what contributes to building a strong change management muscle?*

Figure 6.3 shows the outcomes of that polling. It is no surprise that cultural transparency got the highest score 43% as mastering experiencing in the future has to be built in this transparency foundation where safety of sharing and exchange of ideas, and flowing creativity prevail.

> **Tip**
> Mastery is an excellence journey. Becoming experience-driven is a commitment and balance across management, the selected processes, and the supporting technologies.

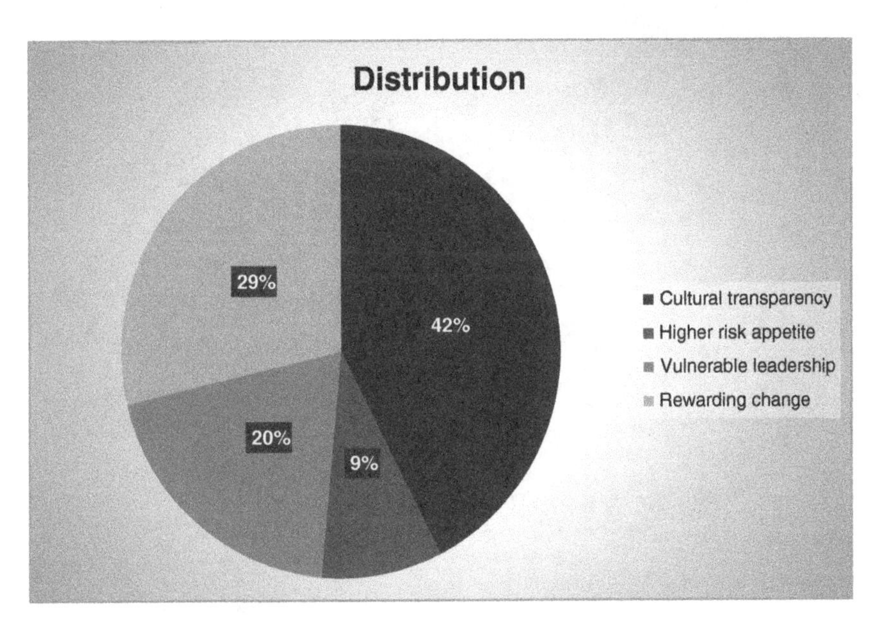

Figure 6.3 Change Management Muscle. *Note*: Based on LinkedIn Open Polling, April 2024.

6.5 The Inspiring Leader

With the massive potential of data and AI, there is one way where leaders could continue to create unprecedented impact. Be inspiring leaders. This is not a place that is likely for machines. Humans connect! The power of human-to-human could only multiply in the future of experiencing. Figure 6.4 shows the potential qualities of an inspiring leader in the ability to connect and drive transformational change. A mix of ingredients contributes to the creation of that leader. Most of these qualities could be learned and developed. A good mix of being attentive, determined, and nurturing, among others, combine to create that view of the leader who inspires us to follow a vision for a changed future state.

One other critical sign of the authenticity behind those inspiring leaders is how humble they are. One of the easiest types of evidence of this is the leaders' ability to surround themselves with others who are more talented than them. In our article, Kerzner and Zeitoun (2024), we address this future view of impactful leaders.

Figure 6.4 Inspiring Leadership. Credit: johnhain/Pixabay.

6.5.1 Introduction

In an era where the focus of many global organizations shifts to handling external disruptions, and with digital being at the core of most of the key ones, the future is likely shaped by the shifts in how we lead. In our research and writing work in the book "Project Management Next Generation: The Pillars for Organizational Excellence," we dedicated one chapter and a pillar to leadership, yet most of all the other 9 pillars address critical points related to leading or ways of working into the future. The definition of what leadership is and what good leadership looks like will likely become one of the frequently changing topics of our times. With the increasing number of external disruptions, the role of the leaders could very well become the one unique differentiator of how value is best created in the work of tomorrow and how effectiveness in decision-making is maintained.

In this article, we tackled a few elements to start the dialogue on what "Inspiring Leadership" could mean and why it is critical to drive and motivate future generations, including the youngest people alive today, generation Alpha. Leadership of the future is more about the art of connectedness. It is the true act of translating context into something that is relatable to ignite the focus of work teams of the future.

If you ask a project manager today, "Do you work for the team members, or do the team members work for you?" the answer would better be that the PM works for the team members and must engage them properly and build trust and strong bonds. We believe this is true servant leadership. Years ago, the PMs would respond that the team works for the PM. Inspiring leaders lean in with this mindset and exhibit a few additional critical attributes to handle turbulence in the future and deliver innovative outcomes.

Excellence of leading in the future is adaptability to the human needs and the next generations' ways of working. We will highlight a few foundational elements from the Next Generation work and use that as the foundation to build the core elements of future inspirational leadership competencies. Simply stated, leadership matters more than ever and true project excellence starts and ends with the fitting leadership principles.

6.5.2 Servant Leadership Matters

Since the 1970s, considerable research has been performed related to the link between controlling emotional labor and servant leadership. Spears [2002] elaborated on Greenleaf's work by identifying ten characteristics of a servant leader:

- **Listening**: Listening is a willingness to openly accept the ideas, opinions, and suggestions of workers.

- ***Empathy***: Empathy extends listening when leaders can put themselves in the situation that others say they are in and empathize with them and their feelings. This is accepting people for who they are.
- **Healing**: The ability of a leader to help workers endure the disappointment and emotional pain that comes from broken dreams, hopes, and other challenges.
- **Awareness**: The ability of the leader to identify cues and signs in the environment to help workers perform better.
- **Persuasion**: Persuasion or persuasion mapping enables the leader to identify the needs of the workers and focus on the importance of their work without the use of formal authority or legitimate power.
- **Conceptualization**: The ability of the leader to think about the future rather than just present-day needs and to encourage workers to use mental models to expand the creativity processes.
- **Foresight**: This includes using intuition to anticipate and predict the future for the benefit of the workers and the organization.
- **Stewardship**: Stewardship involves preparing the organization and its members for great contributions to society thereby willing to serve others.[2]
- **Growth**: Working with team members, possibly on a one-on-one basis, to get them motivated. This, in turn, can lead to employee satisfaction, and the worker is encouraged to perform extra work.
- **Community Building**: Encouraging the workers to view the organization and the team as a community where workers communicate with each other to address their issues.

The ten characteristics opened the door for empirical studies and volumes of literature on servant leadership theory. Some papers discuss only a few of these characteristics. Barbuto and Wheeler [2006], addressed five characteristics, namely altruistic calling, emotional healing, wisdom, persuasive mapping, and organizational stewardship.

6.5.3 Crisis Leadership is a Difference Maker

Crisis leadership requires an examination of the processes that are essential for an organization and its management when dealing with crises. Even though many of these processes, and the accompanying tools and techniques, are based upon best practices and lessons learned from experience, they may not be applicable to crises-related projects without some modifications.

How companies respond to the crisis is critical. Thanks to usually excessive media coverage, the world watches how companies respond to a crisis. Based on

2 For an example of stewardship, see "Why Social Impact Matters", *Pulse of the Profession*® In-Depth Report, 2020, The Project Management Institute, Newtown Square, PA.

the outcome, the public then categorizes the company as either a victim or a villain in the way the crisis was managed. What is expected to be discussed in journal articles will be the project management processes that were used and the accompanying leadership styles.

Most companies today capture best practices and lessons learned from projects during execution and at closure. The best practices look at what the company may have done right and wrong. However, what has been lacking until recently, thanks largely to the pandemic, is a detailed look at the effectiveness of the leadership style that was used and how team members responded. A more in-depth look specifically at crisis leadership can give companies guidance on what type of individuals are best suited to manage crisis projects in the future.

Project managers have become accustomed to managing within a structured process such as an enterprise project management methodology. The statement of work may have gone through several iterations and is now clearly defined. A work breakdown structure exists, and everyone understands their roles and responsibilities as defined in the responsibility assignment matrix (RAM). All of this took time to do.

This is the environment we all take for granted. But now let us change the scenario a bit. The president of the company calls you into his office and informs you that several people have just died using one of your company's products. You are being placed in charge of this crisis project. The lobby of the building is swamped with the news media, all of whom want to talk to you to hear your plan for addressing the crisis. The president informs you that the media knows you have been assigned as the project manager, and that a news conference has been set up for one hour from now. The president also asserts that he wants to see your plan for managing the crisis no later than 10:00 p.m. this evening. Where do you begin? What should you do first? Time is now an extremely inflexible constraint rather than merely a constraint that may be able to be changed. Time does not exist to perform all of the activities you are accustomed to doing. You may need to make hundreds if not thousands of decisions quickly, and many of these are decisions you never thought that you would have to make. This is crisis project management. What leadership style is best for this type of environment?

Historically, many companies were poor at understanding risk management, especially at evaluation of early warning signs. Today, project managers are trained in the concepts of risk management, but specifically related to the management of the project, or with the development of the product. Once the product is commercialized, the most serious early warning indicators can appear and, by that time, the project manager may be reassigned to another project. Someone else must then evaluate the early warning signs.

Early warning signs are indicators of potential risks. Time and money are a necessity for evaluation of these indicators, which preclude the ability to evaluate all risks. Therefore, companies must be selective in the risks they consider.

Future leaders will practice crisis leadership fluidly.

6.5.4 Inspirational Leadership

Looking into the future of inspirational leadership, we will highlight a set of attributes that paint the picture of those leaders. Figure 6.5 highlights eight competencies that will help future leaders connect better and drive transformational change in their organizations. These competencies are drivers to the future project culture.

Purpose clarity is the first of those competencies. Future leaders will put a much higher emphasis on leading with purpose. These leaders will apply the concept of "slowing down to go faster." Taking time to ensure clarity of purpose and that the project teams fully comprehend the reasoning behind their projects, is a critical leadership muscle to build. This will enable cutting losses and increasing efficiencies associated with many of the future transformation initiatives.

The second competency has to do with ***transformational skills***. This is central to the project's way of working and the recognition that projects and programs are about creating change. Transformation competencies build on resilience and adapting and a strong sense of the aspirational potential of initiatives. This connects nicely to the purpose clarity and enables the leader to be consistent in the drive toward transformation outcomes. The 3rd is about **leading with speed** as exemplified by the decision-making process. This is not sacrificing the quality of

Figure 6.5 The Inspirational Leadership Competencies.

the decisions, yet it is about taking more valuable risks and handling uncertainty well. This is an attractive quality of future leaders that connects them well to the inspirational effects they intend to create on stakeholders.

Change management as the 4th is a central competency that future leaders have to master. The change journey that stakeholders follow the leader on, will not always be a comfortable ride. People will be at different stages of readiness for change or willingness to take that on. Leaders have to operate with heart in order to inspire change. ***Empathy*** is a nice 5th and complementary competency that allows the leader to relate well to the future generations, what they need, how they best work, what motivates them, and most importantly how to keep them energized and connected to the mission.

Value focus is the 6th unique characteristic and it goes full circle back to purpose. This is becoming one of the most important focus areas for leaders. Bringing projects back to value and redirecting teams' efforts toward value is a leadership quality that continually brings attention back to the "so what." In a world that continuously deals with distractions and gets overwhelmed with action focus, it is critical that value stays at the center of how leaders focus and prioritize their decision-making process. The 7th is ***communication skills*** which is probably the most coveted characteristic in the topic of leadership. It is amazing how much more potential this competency has. Intentionality of that communication toward inspiring future generations is growing in importance. The clarity of the communication, its design for impact, and its connection to purpose, all come together to enable leaders to inspire change-making.

The 8th and integrating characteristic is ***program management***. This attribute of future inspiring leaders is a must-have. Program management is an integrated value system. This is the strategy connecting competency where programs can be the vehicle for leaders to engage future talent in change initiatives that matter.

While all these competencies are important individually, integrating these in the style of leading in the future brings the inspiring impact of leading to the next heights of impact.

6.5.5 The Path Forward

Shift in how we lead is a must. In future where projects are becoming the norm and where digitization will change everything we do, how we live, how we work, and how fast one ingredient remains intact, the ***human ingredient***. Investing in creating the growth mindset that leaders need to build these 8 competencies, is a good starting point. It is critical however to stay open to any changes to these buckets of competencies. What inspires today's generations and how they work into the future, will likely be disrupted on the path ahead.

This path forward requires boldness and awareness in leading. It assumes that by having a growth mindset, we will continue to learn different ways of leading and adapting to the changing dynamics in how we work and live. This is another way the *Next Gen* leaders would continue to inspire and create the most impactful future transformational changes.

> **Tip**
> Inspirational leadership matters and can be developed. It requires empathizing at a higher level toward communicating and connecting intentionally across stakeholders.

References

Kerzner, H. and Zeitoun, A. (2023). Cracking the excellence code, the great project management accelerator, series article, *PM World Journal*, Vol. XII, Issue XI.

Kerzner, H. and Zeitoun, A. (2024). Inspiring leadership, the future project culture, *PM World Journal*, Vol. XIII, Issue II.

Spears, L. (2002). Tracing the Past, Present, and Future of Servant-Leadership. In L. Spear, & M. Lawrence (Eds.), *Focus on Leadership: Developments in Theory and Research* (pp. 1–16). New York: Palgrave Macmillan.

Barbuto, John E. and Wheeler, Daniel W. (2006). Scale Development and Construct Clarification of Servant Leadership. *Group and Organization Management*, Vol. 31, Issue 3, pp. 300–326.

Review Questions

Parentheses ( ) are used for Multiple Choice, when one answer is correct. Brackets [ ] are used for Multiple Answer, when many answers are correct.

1 What is part of the commitment of an experience-driven culture?

( ) Central-control governance to achieve uniformity.
( ) Full automation commitment.
( ) A strategic way of working that is built around stronger strategic insights.
( ) One size fits all processes.

2 What are possible ways to describe the value of the power of bridging? Choose all that apply.

[ ] Strengthening the horizontal working muscles.
[ ] Project leaders have much more time on their hands.
[ ] Maximizing the use of technology.
[ ] Ability to collaborate.

3 What is the likely style of leadership exhibited by Janet in the case study?
( ) Servant.
( ) Inspiring.
( ) Autocratic.
( ) Focused.

4 What is fundamental way to sustain communications' impact?
( ) Put more emphasis on trends analysis.
( ) It is not necessary to be intentional.
( ) Breaking down bubbles that limit understanding.
( ) Continuously adjusting communications tools.

5 What are useful building ingredients for a stronger change management muscle? Choose all that apply.
[ ] Rewarding change.
[ ] Protecting the team from scope creep.
[ ] Cultural transparency.
[ ] Central leadership.

6 What is a key sign of an inspiring leader?
( ) Deliverables focus.
( ) Purpose clarity.
( ) Walks on water.
( ) Spending more time running the business.

7 What is an important quality in sustaining portfolio value?
( ) Encourage adding more projects to the mix.
( ) Balancing capacity and capability.
( ) Keep investing in trouble projects.
( ) Ensure leaders are certified PfMPs.

7

Digital Fluency

The future is both digital and human. Leaders and team members will not be disrupted by artificial intelligence (AI); they will be distributed by their attitude toward the digital world. This means humans have the keys necessary to develop digital fluency to put them at the helm of effectively operating in the future of work. Only if humans don't adapt or take on the responsibility of developing this new mindset required, then they could be negatively disrupted.

Digital fluency is not just about technology and the capability to use digital platforms. In the realm of experience-driven cultures, it is about the comprehensive understanding of how to be flawlessly connected across the ecosystem. Creating an effective interconnected world also requires that leaders become continually better in relating to contexts and using strategic insights. They are the true strategists and thinkers that experiencing requires.

Key Learnings

- Understand the various sides and edges AI brings to our world and what are the ways to possibly adapt.
- Learn how to create empowering digital experiences.
- Explore the principles of continual innovating and excelling in enhancing speed and scale of innovation.
- Understand the developing attributes of the new human.
- Develop the thinking culture that could balance the investment in humans and technology with a link to clear strategic portfolio outcomes.

7.1 AI Has Many Edges

The world has been preparing for this moment. Artificial intelligence (AI) has many edges. We could choose to focus on the edge where we believe that machines are taking over, or we could choose to capitalize on this moment and usher in a

Creating Experience-Driven Organizational Culture: How to Drive Transformative Change with Project and Portfolio Management, First Edition. Al Zeitoun.

new partnering era. This latter view would be seeing AI as the powerful ally that it could be, and building a close partnership in a widely expanding open ecosystem of partners. Namely, the future.

In an experience-driven culture, this means expedited learning and adapting. Portfolio leaders should refine strategics and strategic choices to ensure optimum utilization of AI capabilities. Developing this AI expertise across program and project teams and the wider business is an executive management responsibility. The platforms that will enable AI adoption are going to continuously become more available and increasingly responsible and ethical.

In the world of portfolios of projects and programs, this directly means a new playing field in the decision-making power for the organization and the teams. The culture of inclusion that has been highlighted earlier in this work will be enhanced by data-powered strategies and the shift of focus that leaders will have toward the positive edge of AI. The leadership landscape will continue to look vastly different in the future, and organizations will have to redesign themselves and take this moment as an opportunity to rethink the talent, the mix of people, and the direct impact of AI capabilities on refreshed business models going forward.

AI tools cover many aspects of project work and ways of running the business. Areas covered span from large language models (LLMs), to design, research, marketing, social media, and marketing, among many others. The infinite mind will be critical to leading in this future of work as training and learning of current AI tools and practices could become obsolete fast. In an AI-driven era, the sponge-like leaning muscles could vastly help leaders be more successful.

On the short-term horizon, hundreds of the world's most powerful companies will continue to experiment with use cases to transform business outcomes and operations. Generative AI agents are a critical part of the leading of projects in the future. This is especially critical given the prediction capabilities that make the trade-offs and the alteration in a project course, much easier and the assurance of achieving outcomes rapidly increasing. Humans could then focus on mastering driving the right strategics, sensing differently, and connecting for impact.

Figure 7.1 is a good illustration of the key message in this part of this chapter. The future of leading requires a highly integrated mix of both machines and humans. It is a true collapse of the real and digital worlds. It is a future where humans are able to stretch to the edge and possibly become a better version of themselves, given where they are able to spend their time. If we own experimenting as a core for where we dedicate time, exercise our brain powers, and openly collaborate, we should be able to come up with more effective use cases, expedite product launches, tackle the most complex global sustainability challenges, and elevate our world.

Figure 7.1 AI Edges. Credit: geralt/Pixabay.

> **Tip**
> Creating the right experimenting focus to utilize the positive edge of AI is a strategic priority. With AI capabilities, growing organizational impact is at a crossroads.

7.2 The Digital Experience

Digital experiences matter in a world that is becoming highly digital. This is especially of utmost importance when we are able to responsibly use digital to educate our ways of working and get the most of our highly valued capital, the humans. Achieving more targeted experiencing at work and across transformational projects makes it fun again to work and show up in our project teams.

The following case study highlights what happens when organizations make structure-changing decisions and attempt to manage portfolios of different types of projects without a clear data-driven strategy. The case likely highlights what not to do in the future of work. The digital experiencing future would allow us to better test operating model change implications, the proper categorization of projects, the selection of fitting leading talent, enhancing trade-offs across customers' request, and develop an intelligent decision-making muscle that minimizes the managing by where the influencing powers exist organizationally.

Case Study

Quasar Communications, Inc.[1]

Quasar Communications, Inc. (QCI) is a 30-year-old, US$ 350 million division of Communication Systems International, the world's largest communications company. QCI employs about 340 people, of whom more than 200 are engineers. Ever since the company was founded 30 years ago, engineers have held every major position within the company, including president and vice president. The vice president for accounting and finance, for example, has an electrical engineering degree from Purdue and a master's degree in business administration from Harvard.

Until 1996, QCI was a traditional organization where everything flowed up and down. In 1996, QCI hired a major consulting company to come in and train all personnel in project management (PM). Because of the reluctance of the line managers to accept formalized PM, QCI adopted an informal, fragmented PM structure where the project managers had lots of responsibility but very little authority. The line managers were still running the show.

By 1999, QCI had grown to a point where the majority of its business base revolved around 12 large customers and 30–40 small customers. The time had come to create a separate line organization for project managers, where each individual could be shown a career path in the company and the company could benefit by creating a body of planners and managers dedicated to project completion. The PM group was headed up by a vice president and included the following full-time personnel:

- Four individuals to handle the 12 large customers.
- Five individuals for the 30–40 small customers.
- Three individuals for research and development (R&D) projects.
- One individual for capital equipment projects.

The nine customer project managers were expected to handle two to three projects at one time if necessary. However, because customer requests usually did not come in at the same time, it was anticipated that each project manager would usually handle only one project at a time. The R&D and capital equipment project managers were expected to handle several projects at once.

1 Kerzner, 2022/John Wiley & Sons.

In addition to the abovementioned personnel, the company also maintained a staff of four product managers who controlled the profitable off-the-shelf product lines. The product managers reported to the vice president of marketing and sales.

In October 1999, the vice president for PM decided to take a more active role in the problems that project managers were having and held counseling sessions with each project manager. The following major problem areas were discovered.

R&D Project Management

Project Manager: "My biggest problem is working with these diverse groups that aren't sure what they want. My job is to develop new products that can be introduced into the marketplace. I have to work with engineering, marketing, product management, manufacturing, quality assurance, finance, and accounting. Everyone wants a detailed schedule and product cost breakdown. How can I do that when we aren't even sure what the end item will look like or what materials are needed? Last month I prepared a detailed schedule for the development of a new product, assuming that everything would go according to the plan. I worked with the R&D engineering group to establish what we considered to be a realistic milestone. Marketing pushed the milestone to the left because they wanted the product to be introduced into the marketplace earlier. Manufacturing then pushed the milestone to the right, claiming that they would need more time to verify the engineering specifications. Finance and accounting then pushed the milestone to the left, asserting that management wanted a quicker return on investment. Now how can I make all of the groups happy?"

Vice President: "Whom do you have the biggest problems with?"

Project Manager: "That's easy—marketing! Every week marketing gets a copy of the project status report and decides whether to cancel the project. Several times marketing has canceled projects without even discussing it with me, and I'm supposed to be the project leader."

Vice President: "Marketing is in the best position to cancel projects because they have inside information on profitability, risk, return on investment, and competitive environment."

Project Manager: "The situation that we're in now makes it impossible for the project manager to be dedicated to a project where he does not have all of the information at hand. Perhaps we should either have the R&D project

(Continued)

Case Study (Continued)

managers report to someone in marketing or have the marketing group provide additional information to the project managers."

Small-Customer Project Management

Project Manager: "I find it virtually impossible to be dedicated to and effectively manage three projects that have priorities that are not reasonably close. My low-priority customer always suffers. And even if I try to give all of my customers equal status, I do not know how to organize my work and have effective time management on several projects."

Project Manager: "Why is it that the big projects carry all of the weight and the smaller ones suffer?"

Project Manager: "Several of my projects are so small that they stay in one functional department. When that happens, the line manager feels that he is the true project manager operating in a vertical environment. On one of my projects, I found that a line manager had promised the customer that additional tests would be run. This additional testing was not priced out as part of the original statement of work (SOW). On another project, the line manager made certain remarks about the technical requirements of the project. The customer assumed that the line manager's remarks reflected company policy. Our line managers don't realize that only the project manager can make commitments on resources to the customer as well as on company policy. I know this can happen on large projects as well, but it is more pronounced on small projects."

Large-Customer Project Management

Project Manager: "Those of us who manage the large projects are also marketing personnel, and, occasionally, we are the ones who bring in the work. Yet everyone appears to be our superior. Marketing always looks down on us, and when we bring in a large contract, marketing just looks down on us as if we're riding their coattails or as if we were just lucky. The engineering group outranks us because all managers and executives are promoted from there. Those guys never live up to commitments. Last month I sent an inflammatory memo to a line manager because of his poor response to my requests. Now I get no support at all from him. This doesn't happen all of the time, but when it does, it's frustrating."

Project Manager: "On large projects, how do we, the project managers, know when the project is in trouble? How do we decide when the project will fail? Some of our large projects are total disasters and should fail, but management comes to the rescue and pulls the best resources off of the good projects to cure the ailing projects. We then end up with six marginal projects and one partial catastrophe as opposed to six excellent projects and one failure. Why don't we just let the bad projects fail?"

Vice President: "We have to keep up our image for our customers. In most other companies, performance is sacrificed in order to meet time and cost. Here at QCI, with our professional integrity at stake, our engineers are willing to sacrifice time and cost in order to meet specifications. Several of our customers come to us because of this. Last year we had a project where, at the scheduled project termination date, engineering was able to satisfy only 75% of the customer's performance specifications.

The project manager showed the results to the customer, and the customer decided to change his specification requirements to agree with the product that we designed. Our engineering people thought that this was a slap in the face and refused to sign off the engineering drawings. The problem went all the way up to the president for resolution. The final result was that the customer would give us additional few months, if we would spend our own money to try to meet the original specification. It cost us a bundle, but we did it because our integrity and professional reputation were at stake."

Capital Equipment Project Management

Project Manager: "My biggest complaint is with this new priority scheduling computer package we're supposedly considering to install. The way I understand it, the computer program will establish priorities for all of the projects in-house, based on the feasibility study, cost-benefit analysis, and return on investment. Somehow, I feel as though my projects will always be the lowest priority, and I'll never be able to get sufficient functional resources."

Project Manager: "Every time I lay out a reasonable schedule for one of our capital equipment projects, a problem occurs in the manufacturing area and the functional employees are always pulled off of my project to assist manufacturing. And, now I have to explain to everyone why I'm behind schedule. Why am I always the one to suffer?"

The vice president carefully weighed the remarks of his project managers. Now came the difficult part. What, if anything, could the vice president do to amend the situation given the current organizational environment?

(Continued)

Case Study (Continued)

Questions

1. Can 13 project managers be controlled and supervised effectively by one vice president?
2. Can the 13 managers under this vice president work effectively with the four product managers under the vice president of marketing/sales?
3. Why does the R&D project manager have built-in conflicts?
4. Should marketing have R&D project managers reporting to them?
5. What are the major problems with small-customer PM?
6. Should the project manager on large projects be permitted to perform marketing activities?
7. Should a company be willing to let some large projects fail?
8. Is it possible for a company to have such a strong technical community that technical integrity is more important than the project itself?
9. Is it possible that capital equipment projects almost always take a backseat to other projects?
10. What specific problems appear in the management of large projects?
11. What specific problems appear in the management of R&D projects?
12. Are there any strengths in the current QCI organization?
13. What type of PM structure is QCI using?
14. What possible recommendation could you make?

Reflections: Having had the vision to invest in digital experiencing, the executive team would have been at a different spot in their organizational maturity.

Leaders and project managers would have been using a different language in reporting against their current top challenges. Decisions on next priority changes would have been much easier to take on. The room for collaboration among the engineers, project managers, marketing leaders, and other executives would have focused on finding more winning possibilities for forward movement and a focus on enhancing a unified customer experience connected to portfolio decision-making.

Tip
Investing in standing up a valuable digital experience contributes to a better understanding of conflicts' context. This empowers leaders to focus on the customer.

7.3 Innovating with Intelligence

In the book "Blitzscaling: The Lightning-Fast Path to Building Massively Valuable Companies" by Reid Hoffman and Chris Yeh, they introduce a strategy for rapidly scaling startups that supports this notion of innovating with intelligence.

The authors address many of the principles that are foundational to the experience-driven focus of this work. Just as in the new mindset targeted in this work and that will lead to creating the new human in the next part of this chapter, they push for cultivating a blitzscaling mindset. The authors include important transformational points such as comfort with risk-taking, embracing uncertainty, and being relentless in getting to the long-term view of impact creation.

That book also aligns with the key points highlighted earlier in this work around experiencing fast, which will help with scaling even before innovation teams are fully ready. This discomfort could have been part of the reason why many of the great successful scaling stories, like YouTube, were possible to achieve. In that case, YouTube was able to leverage the size of networks and infrastructure readiness to seize the moment and scale. The authors equally prioritize culture and values as has been addressed in previous chapters of this work. They view the importance of this as a critical way to manage organizational chaos and operational inefficiencies.

The most relatable aspect in their work to what is critical to innovating with intelligence is the ability to iterate and adapt. Just as emphasized earlier and repeatedly in this work and in structurally executing portfolios of projects, the notion of one-size-fits-all is not where the future is heading. Project teams, just like startups, should be able to pivot, course-correct, and use effective experiencing to become more intelligent in discovery and decision-making process.

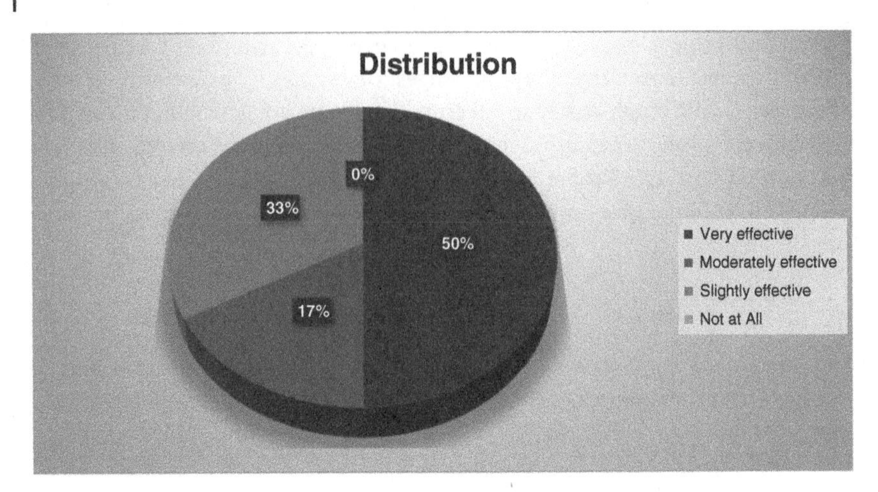

Figure 7.2 Empowering the Human Experience. *Note*: Based on LinkedIn Open Polling, April 2024.

When testing one of the eight hypotheses surrounding this work around digitalization, the views were not as conclusive as one would have except pertaining to the value of digital in enhancing the innovation intelligence. The hypothesis tested was:

Digitalization will empower powerful human experiences

The question used was: *"How effective will digitalization be in empowering powerful human experiences?"*

Figure 7.2 highlights how the answers to this poll were spread. They naturally indicate a high percentage believing in the value of effective digitalization, yet there is still a relatively high percentage (33%) assuming that digitalization will only be slightly effective in empowering the human experience and the possible link to intelligent innovation.

One could assume that has to do with the remaining angst around AI's uncertain edges and the true cultural readiness to implement responsible digitalization on the path forward. It is a transformation room and an opportunity to open the dialogue on how to affect this and how to align our data strategy to the top strategic choices we make, especially for the long-term horizon.

> **Tip**
> Innovating with intelligence builds on the many qualities previously addressed in this work. Most critical will be our ability to iterate and digitally adapt.

7.4 Creating the New Human

Before tackling the building blocks of the new human, let's review the work published on the correlation between digital and one of the more complex dimensions in the planning and managing of projects, namely estimating.

In Kerzner and Zeitoun (2022), we tackle the topic of estimating that creates pain points for the leader and that could potentially benefit tremendously as we power this with AI capabilities in the future. This creates another dimension of the new human, namely a human who spends less time on coming up with an estimate and instead spends more time on steering the achievement of project value across stakeholders.

7.4.1 Introduction

Every year, more people graduate from college and enter the field of PM. Most of these people learn modern PM practices in the classroom but may be unfamiliar with the developments in PM and the problems that modern estimating and PM practices are trying to resolve. To understand the changes that have been taking place and the reasons why certain topics are frequently discussed in periodicals, one must understand the issues we faced initially and how we tried to resolve them, in many cases unsuccessfully until now. For the remainder of this chapter, we will reflect on the topic of estimating and associated data sources and systems.

This is one of the topics that directly contributes to the success of organizational planning efforts. The changes that have been taking place in business and in the way of working of programs/projects have led to an unprecedented level of uncertainty that makes the topic of estimating and the associated risks central to the success of the strategic initiatives.

In this chapter, we also discuss several aspects of the information warehousing growth that drive companies toward the consistent application of business intelligence (BI) systems. It is in our view that digitally enabled estimating requires innovation in order to create a commercially successful product, which also means that the team members must understand the knowledge needed in the commercialization life cycle starting from the early projects' stages.

7.4.2 The Estimating Challenges

During the past 20 years, there has been a significant growth in research surrounding effective PM and estimating techniques. Most of the research focused on functional-related estimating and how to build on the expertise and knowledge base of organizations. The tendency across organizations is to standardize and try to achieve a methodology or a framework that all could follow across organizational verticals. The use of the one-size-fits-all methodology became common practice for many companies for perhaps more than two decades.

What many people failed to realize, either intentionally or unintentionally, was the type of projects that were "forced" to use the methodology. Projects with well-defined requirements and well-written business cases, whether prepared by the client or the contractor, could be successfully estimated, planned, and executed using the one-size-fits-all approach. These were considered as traditional projects. But what about the growing percentage of nontraditional types of projects that may not be well-defined, such as innovation, digital transformation, R&D, and business strategy initiatives? These projects may be initiated based just upon an idea.

There were several issues that began to surface regarding these nontraditional projects:

- The nontraditional project had a much greater impact on long-term competitiveness and profitability than did the traditional projects, but data did not exist to support effective estimating.
- Many of the decisions made by the functional managers on the nontraditional projects focused heavily upon short-term profits that could impact the functional manager's year-end bonus.
- Personal agendas and functional unit objectives were becoming more important than the long-term best interests of the organization.
- In some studies, as much as 80% of the nontraditional projects did not deliver part or all of the business benefits and value expected.
- Articles appeared identifying the benefits of using PM and estimating practices, but the majority of the articles focused on traditional rather than nontraditional projects.
- Executives were unable to make informed decisions in a timely manner due to a lack of reliable metrics to support time and cost estimate accuracy.

7.4.3 Overcoming Estimating Challenges

Overcoming these challenges has not been easy, but significant progress has been made. Articles in journals such as the PMWJ, PMI publications, and new textbooks discussing the changes that are taking place have shown the worldwide PM community of practice that effective managing of change can take place. Most of the challenges we faced over the past several decades are now being eliminated as a result of:

- The growth in the use of flexible methodologies such as Agile and Scrum, either independently or in combination with the firm's existing methodologies, has resulted in a much higher success rate for both traditional and nontraditional projects.

- Advancements in estimating and measurement techniques have allowed project teams to plan, measure, and report project progress much more accurately than with just time and cost metrics. Many of the new metrics measure business, strategic, and intangible factors, allowing executives to make better decisions based on evidence and facts rather than just guesses.
- We now have new definitions of project success, supported by some of the new metrics, which include business benefits and business value created rather than just deliverables produced.
- PM cultures are being created based on trust, which supports the critical dialogues needed for enhanced estimating.
- New forms of PM leadership are appearing that maximize worker engagement efforts and make them feel comfortable to speak their mind without retaliation and then contribute freely to the success of the projects' planning efforts.
- Capturing PM best practices and using them for continuous improvement efforts has become a way of life in most companies.
- The knowledge contained in information warehouses, as well as the amount of information and speed of access, provides companies with a source of competitive advantage.

7.4.4 The Need for Knowledge Repositories

The use of a knowledge management system is expected to become a necessity for all future project teams. Project teams should first map out the mission-critical knowledge assets that are needed to support the project's strategic planning. It is critical to determine which knowledge assets to use and exploit. By mapping the knowledge assets, you set boundaries around what the project is designed to do. Unfortunately, the only true value of a knowledge management system is the impact on the business. Simply stated, we must show that the investment in a knowledge management system contributes to a future competitive advantage.

Knowledge management can increase estimating quality, competitiveness, allow for faster decisions and responses to disruptive changes, and rapid adaptation to changes in the environment. Knowledge management access is critical during design thinking. The growth in information has also created a need for cloud computing. Companies are now creating knowledge management systems and knowledge repositories, as shown in Figure 7.3.

Companies invest millions of dollars in developing information warehouses and knowledge management systems. There is a tremendous amount of rich but often complicated data about customers, their likes and dislikes, and buying habits. This knowledge is treated as both tangible and intangible assets. But the hard part is

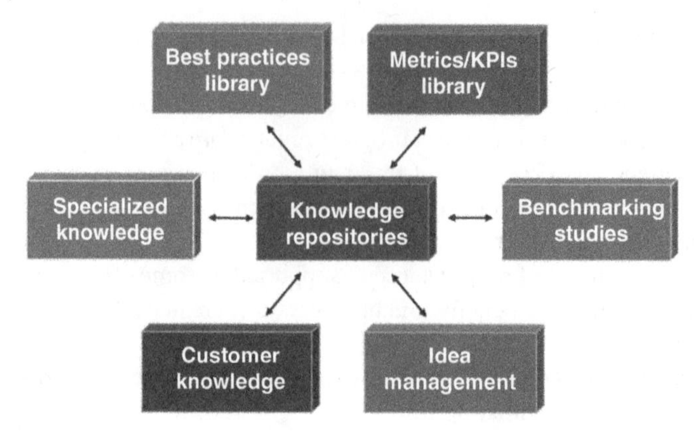

Figure 7.3 Components of a Knowledge Repository.

trying to convert the information into useful knowledge to contribute to excellence in estimating and planning at large.

7.4.5 Intangible Intellectual Capital Assets

The information contained in a knowledge repository is often referred to as intellectual capital. As shown in Figure 7.4, intellectual capital is frequently considered as intangible asset categorized as human, product, and structural capital. These are knowledge-related assets normally not identified on the balance sheets of companies, but they can be transformed into value that leads to a sustainable competitive advantage.

Knowledge databases and information warehouses are needed to support intellectual capital components. These intangible assets that are used to define intellectual capital could be strategically more important to the growth and survival

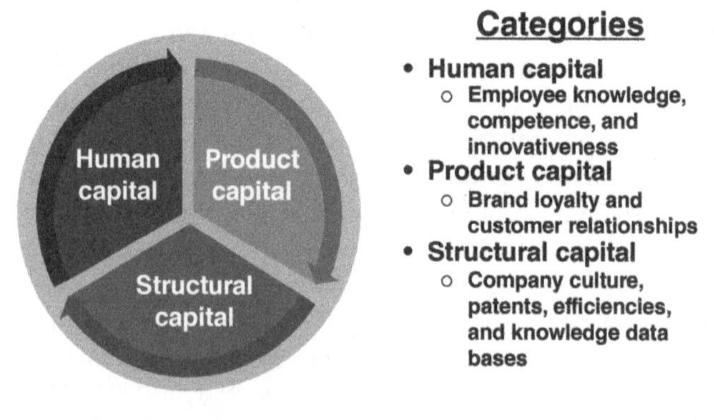

Figure 7.4 Three Critical Intangible Components of Intellectual Capital.

of the firm than its tangible assets. Project teams are becoming more knowledgeable about the importance of intangible assets and are consistently using them to enhance their estimating and planning capabilities.

7.4.6 Categories of Knowledge

In Figure 7.3, we have shown the components of a knowledge repository. The knowledge in each component can come from multiple knowledge sources. There are several sources of knowledge, and they are not mutually exclusive. Table 7.1 lists some ways to classify knowledge sources.

As the future of work, as we continue to highlight in this series, is highly team-centered, project teams must understand their role in driving the use of knowledge assets. Teams should enable their projects to be innovation centers, where continued experimentation takes place and drives future effective estimates that support better digitally empowered planning and decisions.

7.4.7 The Need for Business Intelligence Systems

Simply having knowledge repositories or information warehouses may not be sufficient to support future projects' estimates in an effective manner. BI systems are often considered as the next step after knowledge repositories or information warehouses and combine business information with technologies in a manner that allows project managers to make strategic and/or operational business decisions related to their projects.

The components of a BI system are data gathering, data storage, and knowledge management. Metrics information is a critical component of BI systems. The

Table 7.1 Sources of Knowledge

Source of Knowledge	Description
Explicit	Encoded knowledge that can be found in books, magazines, and other documents
Implicit (or tacit)	Knowledge in the heads of people. Also, knowledge retained by suppliers and vendors. Knowledge may be difficult to explain
Situational	Knowledge related to a specific situation, such as a specific use of a product
Dispersed	Knowledge that is not controlled by a single person
Experience	Knowledge obtained from experiences or observations of clients using the product; must understand user behavior
Procedural	Detailed knowledge on how to do something

information contained in the BI systems can be historical, current, or predictive. The information can come from several sources including strategic and operational PM benchmarking studies conducted by Project Management Offices.

BI technologies are designed to handle large amounts of "big data," whether structured, semi-structured, or unstructured, and present the data on meaningful dashboards so that project teams can make better business decisions and take advantage of business opportunities, especially when managing strategic projects. The technologies used in BI systems allow companies to look at external data (i.e., information from the markets in which the company operates) and internal data (i.e., financial and operational data) together and create BI information to support strategic, tactical, and operational projects. BI systems facilitate corporate estimating and decision support systems by transforming raw data into meaningful and competitive BI. However, there are still companies that believe that BI systems are merely the growth of business reporting systems.

Project managers will need to learn new estimating and decision-making tools including digitalized economics, AI, and the Internet of Things (IoT). With large amounts of data, teams may have to rely upon analytical statistics, which includes:

- **Descriptive data analytics**: analysis of historical data including past successes and failures.
- **Predictive data analytics**: analysis of the data to make predictions of what might happen.
- **Prescriptive data analytics**: look at the reasons why things may happen, estimate options for risk mitigation of future work, and options to take advantage of opportunities.

7.4.8 Big Data

The growth of big data will most likely impact most companies worldwide. For effective analysis of the data, project teams will need workers who possess data science capabilities. The skills will include statistical methods, computational intelligence, and optimization techniques.

There are numerous mathematical models that currently exist to support project estimating and decision-making efforts using big data. A list includes:

- Financial models (Return on Investment [ROI], Internal Rate of Return [IRR], Net Present Value [NPV], payback period, benefit-to-cost ratio, and breakeven analysis)
- Time (scheduling models)
- Money (cash flow models)
- Resources (competency models)
- Materials (procurement models)

- Work hours (estimating models)
- Environmental changes models
- Consumer tastes and demand models
- Inflation effects models
- Unemployment effects models
- Changes in technology models
- Simulation and game models
- Mental models

The expected benefits of using big data effectively include:

- Detection of patterns and trends related to time, cost, and scope.
- Comparison to other projects as well.
- Identification of the root causes of problems.
- Better use of "what if" scenarios.
- Better trade-offs on competing constraints.
- Better tracking of assumptions and constraints.
- Better tracking of Volatility, Uncertainty, Complexity, and Ambiguity (VUCA) and the enterprise environmental factors.
- Better response to out-of-tolerance situations.
- Better capacity planning decisions involving resource utilization.
- Ability to make strategic rather than just operational decisions.
- Ability to make change management decisions.
- Decision-making can be pushed down the organizational hierarchy, but there will be "rules for decision-making" established.
- Emphasis on long-term perspectives rather than just short-term.
- A reduction in the risk of making the wrong estimates or decisions because of a lack of information.

Project teams seem to focus on the knowledge management portion of the BI system. This includes:

- How performance metrics are created and reported
- How benchmarking information can be extracted
- Statistical and predictive analytics
- Data visualization techniques and dashboard design
- Business and project reporting for executives and stakeholders

7.4.9 The Path Forward

Reflecting back on our recent work on the experience culture skills, the project teams of the future will be equipped with new skills such as being data scientists, knowledge asset analysts, and strategically minded leaders. Sensing and responding strengthening are highly data-centered, and strengthening these muscles, focusing on business value, and building higher adaptability to changing customer requirements will enhance estimates and decision quality by the future leaders.

One-size-fits-all estimating or planning models will not exist in the future. The ingredients and building blocks around data warehousing, knowledge assets, and BI will dominate the next decade of estimating capabilities. Leading with data and knowledge-centered objectivity will be a major priority for executives and future leaders.

The path forward requires a strong commitment to the necessary information, tools, and processes to support complex problem analysis and decision-making. Advances in technologies and the growth of information warehouses are driving companies toward consistent application of BI systems.

We believe the future will see a continuation of managing our business by projects and that PM is the delivery system for sustainable business value. Therefore, project managers are expected to deliver better estimates and business decisions, as well as project decisions, and need direct access to a great deal of high-quality project and business information. A digitally enabled and continuous learning-based approach will keep future leaders sensitive and capable of planning, creating, and thriving under tomorrow's disruptions.

7.5 The New Human Attributes

To summarize some of the dimensions of this new human, Figure 7.5 attempts to propose a balanced view of that leader who will be able to chart the course of the experience-driven ways of working ahead of us. The figure shows a few recommended attributes in the categories of head, heart, and hand. The **head** will continue to enhance its intelligence with digital elements, and the **heart** category will continue to adapt the use of language that motivates, connects, and acknowledges human progress.

The **hand** will continue to be at the center of executing with experiencing at its core, thus allowing for the augmentation between the real and the digital toward a potential future of tremendous scaling speed and effectiveness.

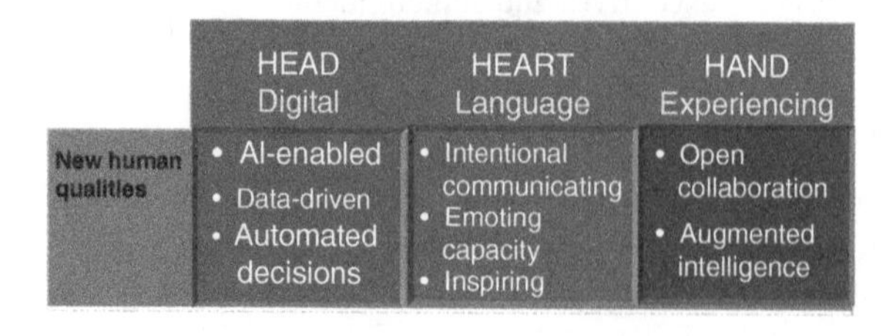

Figure 7.5 The New Human Qualities.

> **Tip**
> Constructing the ***New Human*** is a fine balancing act between head, heart, and hand qualities. An augmented human is a better future version.

7.6 Achieving Balance

One of the most critical dimensions in the experience-driven cultures of the future is the ability to achieve balance. This is the balance between tactical and strategic focus, in addition to the balance between digital and human. It is also about finding the right degree of experimenting in the future and balancing where we spend our time in relationship to effective strategic value realization.

Achieving this balance is a key to future experiencing. Much of achieving the balance topics were also addressed in the interview conducted by PMWJ. The interview was conducted by Yu Yanjuan, PMR (2022). Project Management is a Strategic Competency: Interview with Dr. Harold Kerzner and Dr. Al Zeitoun; *Project Management Review*; republished in the *PM World Journal*, Vol. XI, Issue IV, April.

Q1. Based on your observation, what are the challenges facing project management now? What should we pay attention to in the era of PM 4.0?

Harold Kerzner/Al Zeitoun (Kerzner/Zeitoun): For decades, project management appeared restricted to traditional or operational projects where the requirements were well-defined at the onset of the project. We used to tell students to initiate planning, scheduling, and budgeting activities after they got a scope statement or detailed statement of work. The result was that most of the projects could be executed using a one-size-fits-all methodology. Most traditional or operational projects used the one-size-fits-all approach.

Today, we are seeing new types of projects coming into the mix. These new projects are strategic in nature, such as innovation, research and development, new product development, and strategic planning initiatives. Many of the traditional project management processes, tools, and techniques used in operational projects do not apply to strategic or innovation projects.

Today, executives have realized the value of using project management for all types of projects. Flexible approaches such as Agile and Scrum have been found to be more effective than the traditional waterfall approach on many projects. We find the key here is to not think "either or," but choose a mix that most fits the project's context.

Another challenge is that many strategic and innovation projects start out with an idea, rather than a well-written statement of work or business case. Strategic projects have a greater likelihood of being impacted by even small changes in the enterprise's environmental factors. In addition, decisions in strategic projects entail a higher degree of business risk than with the traditional projects. New metrics will be required to determine the true status and value of strategic and innovation projects. These challenges are now changing the knowledge requirements and the mix of skills that we expect project managers to possess in the future.

Project management (PM) 4.0 is strategic, has a deep understanding of customer needs, and creates a set of principles that must be practiced by forward-thinking organizations that want to grow, adapt, and succeed. PM 4.0 is an area of emerging interest to us because it is well-suited for projects of significant strategic performance for the future of organizations.

Q2. As to how to deal with Volatility, Uncertainty, Complexity and Ambiguity (VUCA), do you have some suggestions for professionals?

Kerzner/Zeitoun: When there is commonality among the projects in a firm such that a one-size-fits-all approach can be used during project execution, the needed skill sets may be known with some degree of certainty. In such cases, there exists a well-defined statement of work (SOW), and the impact of the enterprise's environmental factors is relatively low. But today, where project managers are now responsible for managing strategic projects, new skills are needed to meet the new business challenges.

Today, and in the future, project management will take place in a VUCA (Volatility, Uncertainty, Complexity, and Ambiguity) environment. Strategic projects are more susceptible to the enterprise environmental factors in the VUCA environment than traditional projects, thus requiring the reskilling of project managers.

The meaning of the VUCA components can change from industry to industry, company to company, and possibly project to project. The impact of VUCA can also change the environment in which the project takes place. The enterprise environmental factors in a project can have a serious impact on VUCA analysis and subsequent risk management.

VUCA of a project also impacts the culture of a firm. Strategic projects will vary from company to company, and even in the same company, there can be a multitude of different types of strategic projects included in innovation, R&D, entrepreneurship, new product development, and changes in business models. The skills needed can vary based on the type of strategic project.

As an example, different skills may be needed for strategic projects that demand radical rather than incremental changes in how the firm conducts business. Some of the new skills needed for strategic projects include design thinking, rapid prototype development, crowd-storming, market research, brainstorming, and change management. For project managers involved in multinational strategic projects, the list of skills might also include an understanding of local cultures, values, and politics that are evident during a VUCA analysis.

Effective project management requires not only an understanding of project management and the deliverables expected from the project but also the relationship that the project has with ongoing business activities and strategic planning. VUCA activities add significant risks to all of these relationships. Therefore, risk management—especially business risk management—could be one of the most important skills needed for future project managers. In the past, business risk management related to projects was considered a responsibility of the project sponsor, the project governance committee, and even senior management.

This is no longer the case. Project teams must become more proficient in risk management resulting from the VUCA factors. This mix of skills and shifts required of professionals to deal with this VUCA environment confirms the need for strengthening the muscles of adaptability and resilience.

Q3. Recently, there has been a lot of talk about "Project Economy." It is believed that projects will serve as the engine of the future economy. What are your views on it? It seems that the future of project management is promising, right?

Kerzner/Zeitoun: Today, more companies believe that they are managing their business as a series of projects. Trust in the abilities of project managers has increased significantly. As executives recognize the benefits of utilizing effective project management practices on all types of projects and more trust is placed in the hands of project managers, project managers are being asked to manage strategic projects as well as the traditional or operational projects.

Trust in asking PMs to manage strategic projects has resulted in the establishment of a line-of-sight from project managers to senior management such that project managers are kept informed about strategic business objectives to ensure that strategic projects are aligned correctly. Line-of-sight creates not only a correct decision-making mindset, but it also provides project managers with more knowledge about the company, thus reducing the chance for ineffective behavior. Line-of-sight can also make it easier to develop the proper risk management mindset. The notion that "information is power" is disappearing in the project management landscape as strategic information is being shared.

Many companies today conduct a study every year or two to identify the four or five strategic career paths in the company that must be cultivated so that the growth of the firm is sustainable. Project management makes the short list of these four or five career path slots. As such, project management is now treated as a "strategic competency" rather than just another career path position for the workers.

Part of this is evident by looking at to whom project managers now report project status and make presentations. Historically, PMs conducted briefings for the project sponsors and occasionally senior management. Now, with the responsibility to manage strategic projects that may impact the future of the firm, project managers may be conducting briefings for all executive management and even the board of directors. As this shift from running the business to growing the business continues, the future of project management is certainly promising for decades to come.

Q4. Virtual work is getting increasingly common. What are the challenges and opportunities resulting from virtual work?

Kerzner/Zeitoun: Companies have come to the realization that knowledge needed for sustainable business growth may not reside entirely within their company. Developing global business partnerships provides significant business advantages, such as lowering of project costs, faster time to market, improvements in quality and reliability of products and services, greater customer satisfaction, and lowering of project and business risks.

All of these benefits are achievable as long as the parent company maintains a good grasp of virtual teams. Understanding the benefits is often easier than understanding the challenges. Some of the challenges include time zone differences, limited collaboration opportunities, people being afraid to state their true feelings, and possibly the inability of team members to possess the same comprehensive information as they would if the team were collocated.

Virtual meetings are somewhat more difficult than onsite meetings because the virtual environment requires a different set of tools and software for communication, viewing, recording and displaying of ideas, and interaction among participants. If the group must be broken down into smaller groups, multiple concurrent virtual sessions may be necessary.

Virtual teams have advantages and disadvantages. The benefits include:

- Participants are under less peer pressure and may not be intimidated by others on the call.
- It may be easier to put together a diverse team of participants.
- People are working alone or in small groups and may come up with more fruitful solutions to project problems than in larger groups.
- Large groups can participate virtually, and it is less likely that someone will want to dominate the discussion with their ideas.
- Large groups can be subdivided into smaller groups without worrying about title, rank, or expertise.
- There is less wasted time in virtual sessions than with in-person sessions.

Disadvantages of virtual teams include:

- Facilitators must ensure that the proper virtual tools are in place.
- It may take more time at the onset of the meeting to make sure that everyone is on the same page.
- Sharing documents may be difficult virtually; facilitators must ensure that all participants have the appropriate materials.
- The way that communication takes place may make it difficult for people to build on the ideas of others or to combine ideas.
- It may be difficult to break large groups into smaller groups virtually.
- Virtual participants may be less likely to ask questions than if they were in the room with the other team members.
- Having an open dialogue where everyone gets to speak may be difficult to enforce.
- Having too large a group may prevent or discourage members from providing input.
- People may be multitasking or distracted, and the facilitator has limited control over the meeting.
- Perhaps the biggest challenge in virtual teams is the inability to read body language and, therefore, not fully knowing how others feel. Difficulty in observing facial expressions and nonverbal behavior such as what they do with their hands or the way they are sitting as an indication of whether one is upset or in agreement are examples of body language.

But today, we believe that the advantages significantly outweigh the disadvantages, and virtual teams will continue to grow.

Q5. What are the characteristics of projects in the future?

Kerzner/Zeitoun: There are numerous characteristics that we can predict for projects of the future, but perhaps the greatest characteristic will be the use of flexible methodologies, new metrics, and new leadership styles. Historically, companies used a one-size-fits-all methodology that was inflexible and used the same life cycle phases for every project. Unfortunately, this rigid methodology was not effective on several types of projects. With the growth of flexible methodologies, project teams will be allowed to establish their own life cycle phases and ways of working within reason.

Previously, we stated that project management has become a strategic competency and that most projects are aligned to strategic business objectives. To select and evaluate these new types of projects, there must exist strategic project metrics. We cannot rely entirely on the traditional metrics of time, cost, and scope to determine project status and business impact.

Strategic business metrics must be able to be combined to answer questions that executives and active stakeholders might ask. The list below identifies metrics that executives need to make decisions concerning business and portfolio health.

- Business profitability
- Portfolio health
- Portfolio benefits realization
- Portfolio value achieved
- Portfolio mixture of projects
- Resource availability
- Capacity utilization
- Strategic alignment of projects
- Overall business performance

Project teams must provide input to these metrics for alignment with strategic objectives.

For decades, companies have recognized the existence of intangible assets, but only recently has the importance of measuring intangible assets such as improvements in project governance become important. Measuring intangible assets can improve project performance, and today, the growth in measurement techniques has made this possible. Measuring the growth in intangible assets may be dependent upon management's commitment to the measurement techniques used. Also, the techniques must be free of manipulation.

Examples of intangible assets related to projects include:

- Improvements in goodwill.
- Improvements in customer satisfaction.
- Improvements in our relationships with our customers.
- Improvements in our relationship with our suppliers and distributors.
- Improvements in our brand image and reputation.
- Growth in patents, trademarks, and other intellectual property.
- Effectiveness of the execution of business processes.
- Effectiveness of executive governance.
- The company's culture and mindset, and how they impact projects.
- Growth in human capital, including retained knowledge and the ability to work together.
- The effectiveness of strategic execution and decision-making.

All these assets are measurable. The value of intangibles can have a greater impact on long-term strategic business considerations rather than short-term factors. Management support for the value measurement of intangibles can also prevent short-term financial considerations from dominating project decision-making.

Another challenging topic will be project leadership. Most people seem to agree that effective leadership in project management can contribute significantly to successful outcomes. Unfortunately, there have been limited empirical studies and research on project management leadership styles and their impact on the performance of team members and their present and future assignments. On the other hand, there are volumes of information related to project management processes, methodologies, tools, and techniques.

Project management environments are generally unstable and are likely to change from project to project. Each company, even in the same industry and with similar projects, can operate in different settings based on a variety of factors. Some educators believe that although no definitive leadership style is recommended, project management is closely aligned with situational leadership practices.

New methodologies, such as Agile and Scrum, have a strong focus on collaboration. Project managers will no longer view team members as a cost. Projects are getting longer and more challenging. Project managers must develop leadership skills that engage, secure commitment and participation, and foster an environment where everyone feels safe to express their true opinions and feelings. Social project leadership will replace the traditional autocratic leadership styles that are still being practiced in some organizations.

Q6. What are the skills needed to manage projects in the future? Design thinking, systems thinking, resilience, adaptability, etc.? Please offer your list and a little bit of elaboration on each.

Kerzner/Zeitoun: Given that many of the projects we will manage in the future focus on strategic objective accomplishment and business value rather than traditional deliverables, project teams must learn new skills. Some of these skills include:

Brainstorming: More projects in the future will begin with just an idea rather than a business case and detailed statement of work. Project teams must therefore learn how to participate effectively in brainstorming sessions.

Creative problem-solving: This involves ways of looking at a fresh perspective to solve a critical problem. This requires out-of-the-box thinking.

Design thinking: This is a structured process for exploring ill-defined problems that were not clearly articulated, helping to solve ill-structured situations, and improving outcomes. The focus of design thinking is generally finding solutions to a problem rather than identifying the problem.

Idea management: This involves ways of capturing all ideas for solutions to a problem and retaining all ideas, whether they are used or not, in an information warehouse for future usage.

Rapid prototype development: Techniques for developing rapid prototypes throughout a project to support continuous decision-making are needed, rather than just one prototype near the beginning of commercialization.

Innovation leadership: The ability to provide leadership to projects in the future that may begin with just an idea and require different forms of innovation is essential.

Strategic planning: The ability to make project decisions that must be aligned to strategic business decisions will be key.

Managing diversity: The ability to manage teams, perhaps in a virtual environment, that have team members from various backgrounds is crucial.

Co-creation team management: This is the process of working with external resources that bring knowledge, expertise, and ideas to your project.

Supply chain management: Given that it may be cost-effective to outsource more work than previously done, project teams will take a more active role in supply chain management.

Advanced risk management: Many of the new types of projects will require business and strategic decisions, as well as being impacted by the VUCA environment. This will add more challenges to the team's ability to respond to risks.

Change management: The results of many projects may require critical organizational change management. Project managers will be participating in change management activities.

Q7. How should PMOs transform to adapt to the future needs?

Kerzner/Zeitoun: Project management success is ownership-based. For any of the changes we highlight in this interview to succeed, we need a maturing discipline in the practice of project management. The only logical place for that ownership is a refreshed view of the future PMOs. It is not our goal here to predict what to call these PMOs. All names would work. Whether we talk about an Enterprise PMO, a Strategy Execution Office, or a Global Center of Excellence, these versions could all work if the context is properly understood.

The future context of PMOs is truly empowerment-based. This means that it is no longer a luxury for the PMO to be fully authorized to operate strategically. This will be the only accepted currency in the future. The PMO has to be able to operate objectively and drive most critical strategic dialogues in tomorrow's organizations. How else would the project economy continue to prevail or the supporting teams' autonomy succeed? How can we have assurance that the next useful set of practices will be implemented, or could all critical benefits be realized?

The PMO has to transform. Just like the many changing project management skills highlighted in this interview, the skills of strong collaboration, high adaptability, and strategic thinking are going to drive leaders in future PMOs. The

PMO has to realize that its success hinges on saying "no" much more often than saying "yes" to ensure that strong strategic choices are being consistently made. This also means that enterprise risk management will prevail in how PMOs instill the project management principles across project teams and organizations of the future.

The command in the use of digital will also draw a very different picture for tomorrow's PMOs. Understanding how to use digital as a true enabler will free up time for project professionals, allowing them to step into the strategic role projects and their teams could play. They will also need to address many of the demanding expectations of the VUCA environments and the expanding social and diverse agendas that will prevail in the coming decade.

Q8. What should the profession of project management do to help deal with global crises? David Pells has advocated that the profession of project management should shoulder more social responsibility. Do you agree? Why or why not?

Kerzner/Zeitoun: Companies today maintain a self-regulated strategy called corporate social responsibility (CSR) that is integrated into the firm's business model and identifies the ethically oriented activities the firm will undertake for the benefit of consumers, society, ecology, and government regulations. Included in the description of a firm's CSR is usually the term "sustainability," which may be defined as improving human life without impacting the capacity of the supporting ecosystems. This leads us to the term "sustainable innovation," which is the creation of products and services that support sustainability and CSR. The outcomes of sustainable innovation must support the economy, environment, and society.

However, before a firm invests heavily in sustainable innovation activities, it must understand possible trade-offs between short-term profitability, pressure from investors for a reasonable ROI, and social and environmental goals. Companies believe that, by creating a social value proposition that supports sustainability efforts, they will gain consumer loyalty and trust. Sustainable innovation has traditionally been used in reference to business innovation sustainability, which is the continuous development of new products and services to increase the firm's financial objectives such as market share, revenue, profitability, and shareholder value. However, sustainable innovation can also be aligned with social innovation and environmentally friendly innovation activities that are part of business innovation sustainability.

Social innovation involves more than just improving the quality of life and human well-being. The social activities that the firm can consider as part of CSR include philanthropy, volunteer work, the way it markets and sells its products and services, and the products it creates using natural or renewable resources that do not impact the environment. Environmentally friendly sustainability includes the consumption of certain natural and renewable resources, such as water, energy, and other materials. Environmental innovation activities might force us to consider the impact that our innovation and commercialization decisions can have on global warming, depletion of the ozone layer, land use, and human health considerations resulting from the use of toxic pollutants.

Humanitarian or social innovations are most frequently a subset of public sector innovations but can occur in the private sector as well, based on a firm's commitment to its social responsibility program. We agree that in the expanding project economy, there will be a significant role that projects will play in the sustainability and social responsibility agendas of future organizations. Leaders with a project mindset will be able to govern in a balanced way between the short-term metrics and the long-term socially responsible footprint these organizations leave behind.

Q9. Sustainable project management is a trend. What are your tips on persuading people to think beyond the "Iron Triangle" to consider long-term impacts such as sustainability of the deliverables?

Kerzner/Zeitoun: As we addressed earlier, the classic Iron Triangle has been dismantled with the shift to strategic projects, the heightened focus on value achievement, and the inclusion of social responsibility and sustainability objectives.

No matter how digital the future might be, the human component is a critical ingredient of future views of success. There are organizations that are committed to humanitarian concerns such as the Global Alliance for Humanitarian Innovation (GAHI). The innovations created by these organizations are driven by humanitarian needs, usually involving health and safety concerns, and are designed to save lives and reduce human suffering of vulnerable people. Innovation projects can involve disaster sanitation, saving children and refugees, disease control and reduction, emergency relief possibly from weather concerns, sanitized water, medicines, electricity generation, and refrigeration.

Humanitarian innovations depend heavily on private donors and usually have spokespeople who are well-known actors, actresses, and/or professional athletes. The choice of spokesperson is based on the target audience for the humanitarian innovation, and possibly the geographical area where the spokesperson may be well known and has proven ability to promote donations.

As an example of a humanitarian need, many developing countries suffer from severe acute malnutrition, which is a life-threatening condition that requires urgent treatment. Until recently, severely malnourished children had to receive medical care and a therapeutic diet in a hospital setting. With the advent of ready-to-use therapeutic food (RUTF), large numbers of children who are severely malnourished can now be treated successfully in their communities, which has the potential to transform the lives of millions of malnourished children.

The future of project management is bright if we are able to expand our views from any traditional view of measuring success. The Iron Triangle is past tense. The future tense involves business-, value-, and strategic-driven assessments of the intended success journey of projects and portfolios.

Q10. Throughout the PMBOK Guide (7th edition), creating value is the core theme. Compared with the previous editions, it has changed from process-oriented to principle-based. What's your comment on it? What do you suggest organizations do to transform from plan-driven to value-driven?

Kerzner/Zeitoun: For years, the definition of project success was the creation of project deliverables within the constraints of time, cost, and scope. While this definition seemed relatively easy to use, it created several headaches.

First, companies can always create deliverables within time, cost, and scope, but there is no guarantee that customers will purchase the end results. Second, everyone seemed to agree that there should be a "business" component to project success but was unable to identify how to do it because of the lack of project-related business metrics. Third, this definition of project success was restricted to traditional or operational projects. Functional managers who were responsible for strategic projects were utilizing their own definitions of project success, and many of these strategic projects were being executed under the radar screen because of the competition in the company for funding of strategic projects.

Today, companies believe they are managing their business as a stream of projects, both strategic and traditional. As such, there must exist a definition that satisfies all types of projects. The three components of success today are as follows:

1) The project must provide or at least identify business benefits and value expected;
2) The project's benefits and value must be harvested such that they can be converted into sustainable business value that can be expressed quantitatively; and
3) The projects must be aligned to strategic business objectives.

With these three components as part of the project's success criteria, companies must ask themselves when creating a portfolio of strategic projects, "Why expend resources and work on this project if the intent is not to create sustainable business value?"

These three components can also be used to create failure criteria as to when to pull the plug and stop working on a project. Since these three components are discussed in current PMI literature and PMBOK 7th edition, it is expected that these three components—especially business value creation—will appear in future developments in the standards for project management. The responsibility remains in organizations' hands to select the right mix of processes and principles that enable the achievement of this maturing view of projects' success.

We are now focusing on value-driven project management rather than requirement-driven deliverables. The intent is to create business value, and this will require a different thought process for many professional project managers.

Tip

Achieving balance in the future requires a holistic understanding of the changing dynamics of how projects and work will be conducted and how leaders align on value.

Reference

Kerzner, H. and Zeitoun, A. (2022). The digitally enabled estimating enhancements: the great project management accelerator series. *PM World Journal* XI (VII).

Review Questions

Parentheses () are used for Multiple Choice when one answer is correct. Brackets [ ] are used for Multiple Answers when many answers are correct.

1 What is the edge of AI that enables the experience-driven culture?

( ) Slowing the decision-making process.
( ) Motivating employees by making them anxious about their roles.
( ) The true collapsing of the real and digital worlds.
( ) Cyber security concerns.

2 What were some of the examples of the QCI case that could limit experiencing? Choose all that apply.

[ ] Insisting on a similar approach across portfolios of projects.
[ ] Project leaders have the proper degree of authority.
[ ] Maximizing the use of technology.
[ ] Turf wars across organizational power houses.

3 What is a core principle in blitzscaling that maps most to effective experiencing?

( ) Focus on short-term view.
( ) Scaling only when fully ready.
( ) Iterating and adapting.
( ) Increased governance.

4 What is an example of a benefit of using big data effectively?

( ) Put emphasis on short-term perspective.
( ) Pushing decision-making upwards.
( ) Better trade-offs on competing constraints.
( ) Slower response to out-of-tolerance situations.

5 What are the three intangible components of intellectual capital? Choose all that apply.

[ ] Structural capital.
[ ] Economic capital.
[ ] Human capital.
[ ] Product capital.

6 What is a key sign of a proper use of heart in the new human qualities?

( ) More communications.
( ) Inspiring action.
( ) Automating decisions.
( ) Rewarding individual contributions.

7 What is a disadvantage of virtual teams in the necessary experiencing?

( ) Large groups can participate virtually.
( ) Having an open dialogue where everyone gets to speak may be difficult to enforce.
( ) Less peer pressure.
( ) Smaller groups without worrying about title.

Section III

Creating Experience-Driven Cultures with Enterprise Portfolio Management Muscles

Section Overview

This section is focused on establishing the linkages between the fundamental features behind the experience-driven cultures and the proper exertion of portfolio management practices. Portfolio management, like culture, requires a combination of art and science. There are driving standards, techniques, and best practices that support proper portfolio management across its lifecycle, and when they are implemented within the context of the right supporting culture and the proper inspiring leadership, the effectiveness of these practices increases and directly supports scaling the value of projects' outcomes.

Fostering portfolio management practices supports organizations on their journey to strengthen their strategic execution muscles. Portfolio management discipline should be a strategic priority and a fundamental way of working for the management teams. Many leaders conceptually know what needs to be done to manage a portfolio of work, yet the full and mature implementation of this discipline widely varies across industries and business units within organizations. This is likely affected by the prevailing culture, multiple types of behaviors of the leaders, possible silos, ways of budgeting, approaches to governance, geographies, and varied levels of initiatives' complexities.

Section Learnings

- The relationship between experience-driven culture and successful portfolio management practices.
- The qualities of leading toward excellence in building an ownership culture.
- How do we use an enterprise view of the portfolio to drive enhanced decision-making?

Creating Experience-Driven Organizational Culture: How to Drive Transformative Change with Project and Portfolio Management, First Edition. Al Zeitoun.

- Understanding the critical differentiating skills for mature portfolio management practices.
- Adapting to the changing execution realities and dynamics of future organizations while creating strategic consistency.

Keywords

- Success
- Communication
- Strategy
- Enterprise
- Decision-making
- Value-based
- Portfolio
- Momentum
- Balance

Introduction

The future cultures will benefit from implementing portfolio management practices at the enterprise level. Pockets of organizations could have developed levels of portfolio management maturity, yet connecting this at the enterprise level is a critical integrating step forward.

In this section, this work will continue to tackle the transformational aspects to drive future experience-driven culture success. This will be done by investigating the process side and addressing the portfolio management practices that matter in effective future operations. The leadership side of the equation will also be addressed by reviewing how to develop the necessary resiliency in dealing with the dynamic changes and the multiple unknowns and environmental factors that affect the enterprise portfolio. The infrastructure and other aspects of readiness will also be addressed, with special attention to decision-making importance to effective portfolio management.

The remaining hypotheses of this work will be reviewed, and a few case studies will support the combination of good practices and some of the learning around what not to do in the future. The experts' interviews will complement what is essential in creating these future cultures and connecting their success to portfolio management practice.

8

Building the Experience-Driven Culture

This chapter is focused on the continuation of the steps surrounding the creation of an experience-driven culture with a focus on the execution of its features.

Continuously revisiting what success looks like is a strategic responsibility to ensure that we continue on the path of creating a culture that fits the needs of the organization and its portfolio of projects. Some critical aspects of the ways leadership communicates its vision and how critical elements of the experience-driven culture are cascaded with the right tone will be explored.

Decision-making is one of those areas that typically differentiate organizations and their leaders. Linkages to enterprise risk and the accessibility to the right data to support decisions will be addressed. The transparency created by technology and digitalization will be reviewed in the context of building the right momentum for tomorrow's digital enterprise and sustaining the features and ingredients necessary for fluid experience across program and project teams.

Key Learnings

- The success ingredients for building the experience-driven culture and the overall linkages to portfolio management.
- The importance of setting the right organizational tone and how this reflects on the prevailing behaviors and practices in a given culture and across projects.
- How to improve the quality and speed of decision-making and ensure close ties to value?
- Empowering the organization and putting the right strategy in place to ensure the continuing momentum for maintaining the culture features.

Creating Experience-Driven Organizational Culture: How to Drive Transformative Change with Project and Portfolio Management, First Edition. Al Zeitoun.

8.1 Success Ingredients

Succeeding in implementing the features of experience-driven cultures as highlighted previously in this work, is a commitment that requires continual adjusting to ensure proper fit to what success looks like for a given organization, a given stage of its maturity, and to what strategic value could mean at that given point in time.

Sports, especially team sports, provide a great analogy to the dynamic nature of revisiting success and its fit with the team's culture and the circumstances the team finds itself in. Let's take the example of soccer (or football) and the many parallels it has with culture of organizations and the strategizing that happens toward the successful execution of outcomes.

There are definite elements of strategizing that continue to happen throughout the 90-minute football match. Navigating the various changes and the surprise moves or goals created by the opponent team requires a high degree of flexibility and willingness to change tactics. This will reflect on team dynamics and the ability of the team members to cover weaknesses and gaps they uncover live as they actually go through the execution of the match.

A key ingredient for the success of the team is how well they adapt and make use of the trust and strong relationships they have managed to build among themselves. In addition, the leadership attributes, covered earlier in this work, especially around inspiring leadership, will be tested during those pressure moments when the team is struggling or has fallen behind on the core sheet.

Just as previously addressed around the concept of slowing down to go faster, sometimes this is what the team needs to turn things around in the match and in the final results. This is also why halftime is a great milestone for potential reset. Just like we would encounter in a project or ultimately a portfolio lifecycle, it is important to iterate through journey and keep a clear line of sight for what an updated successful mix of ingredients might be in relationship to the experiencing level needed across the organization and its portfolio of projects. These are critical moments to gauge whether we are still on a relevant path of success and how much the learnings gained, require us to adapt and adjust course.

Figure 8.1 highlights the bonding and connection a football team has and how critical the created connected network of players is in being capable of doing those necessary shifts and ultimately reflecting an experiencing culture. This could be evidenced by changes in the playbook followed or the next moves that the team takes to adjust strategy and final results.

Figure 8.1 Football Team Adapting. Credit: BorgMattisson/Pixabay.

> **Tip**
> Invest in revisiting the definition of success along the way of creating the experiencing culture. Just like in sports, adapting strategy is key to winning.

8.2 Tone Matters

Leading projects in the future are driven by a strong mix of human and digital balance. The human side of the equation determines the level of commitment and decision ownership necessary for successful implementation of the work of a given portfolio of initiatives. Tone is a common intersection word between what is necessary to build the right experiencing culture of the future, and what it takes to inspire and connect team members toward the successful integrated movement on achieving value.

The following case study highlights some of the principles that directly speak to the importance of setting the right tone and what is needed to nurture that tone throughout the lifecycle of a transformation program or as the organization matures its ways of working and operations. In this case, the focus was on where and when across a given portfolio does the senior management team needs to be involved. The tone had to be set around what is considered to be a crisis and the approach they elected to use, was metrics, with targets, and ranges.

Case Study

LXT International[1]

Background

LXT International is a global company that manufactures electronic components. They are one of the leaders in their field mainly because of the innovations that they bring to the marketplace, often yearly.

The competitive nature of the marketplace has forced LXT to make some changes in relation to how they manage the innovation processes. LXT established small innovation units across the world, each unit headed up by a manager with the title "Innovation Office Manager." All the innovation units report to a vice president at corporate entitled "Vice President for Innovation and Growth." The intent of inserting the term "innovation" in their titles was to show that the company recognizes the importance of innovation and that it is part of LXT's culture.

The charter for the innovation units included the following:

- Maintain constant communications with the customers we serve so that they understand how important they are to LXT.
- Determine future strategic needs that our customers have and how our products might satisfy their needs.
- Invite the customers to provide us with their ideas for product enhancements as well as new products.
- Keep the customers informed about the new products we are developing to see if they are interested.
- Allow certain customers, when feasible, to work with our design teams as cocreators.

The innovation units serve as sources of information concerning clients in their geographic region and relay this information to the VP for Innovation. The VP and her team evaluate the information and make the final decision on the allocation of innovation and R&D funds for various projects.

1 Kerzner, 2022/John Wiley & Sons.

The innovation projects fall into three general categories: incremental, radical, and disruptive. Senior management determines the priority of the projects across all categories.

Members of senior management at LXT act as sponsors and provide governance on selected projects. However, because of the large number of projects, many innovation activities are sponsored by middle- and lower-levels of management.

Sponsors and governance personnel usually support the project managers whenever certain decisions are necessary. However, when there exists a significant problem on a project where sponsors need additional assistance, all senior management may get involved. The problem facing senior management was the differentiation between a problem and a crisis. Senior management did not have the time to get involved in all problems but wanted to provide support for problems that were viewed as potential crises.

Defining a Crisis

After conducting research, LXT had a better understanding of what constituted a crisis. A crisis can be defined as any event that can lead to an unstable or dangerous situation affecting the outcome of the project. Crises imply negative consequences that can harm the organization, its stakeholders, and the public. Crises can cause changes to the firm's business strategy, how it interfaces with the enterprise's environmental factors, the firm's social consciousness, and the way it maintains customer satisfaction. A crisis does not necessarily mean that the project will fail, nor does it mean that the project should be terminated. The crisis could simply be that the project's outcome may not occur as expected.

Some crises may appear gradually and can be preceded by early warning signs. These are referred to as "smoldering" crises. Management believed that metrics and dashboards could assist in identifying trends that could indicate a crisis is coming and could provide the project manager with time to develop contingency plans and take corrective action. The earlier people know about an impending crisis, the more options may be available as a remedy.

Another type of crisis is one that occurs abruptly with little or no warning. These are referred to as "sudden" crises. Examples that could impact projects might be elections or political uncertainty in the host country, natural disasters, or the resignation of an employee with critical skills. Metrics and dashboards cannot be created for every possible crisis that could exist on a project. Sudden crises cannot be prevented.

(Continued)

Case Study (Continued)

Not all out-of-tolerance conditions are a crisis. For example, being significantly behind schedule on a software project may be just a problem but not necessarily a crisis. However, if you are behind schedule on the construction of a manufacturing plant and plant workers have already been hired to begin work on a certain date, or the delay in the plant will activate penalty clauses for late delivery of manufactured items for a client, then this could constitute a crisis.

Crisis Dashboards

LXT recognized that there was a difference between risk management and crisis management. Risk management involves assessing potential threats and finding the best ways to avoid those threats. Crisis management involves dealing with threats before, during, and after they have occurred. That is, crisis management is proactive, not merely reactive. If metrics could be developed for potential crisis identification, then LXT would have an early warning system to deal with potential crises.

Another issue was determining how a crisis could be identified using metrics on a crisis dashboard. In an ideal situation, senior management would turn on their computers at the start of the day and open the crisis dashboard. On the screen would appear metrics and information related to potential crises on certain projects. These would then be the only projects that the executives would see on their screen and need to interface with on that day.

The problem was how to identify when a metric was in trouble given that each project could have 20 or more metrics. Management understood that hitting a target exactly requires some degree of luck. Missing a target may be acceptable if the variance from the target, whether favorable or unfavorable, falls within acceptable limits. The limits are referred to as the tolerances, thresholds, or integrity of the target. Therefore, when establishing a target for each metric, it is important also to establish the limits. The established limits must be acceptable to the project team, the client, and the stakeholders. Typical limits might be the target $\pm 5\%$ or the target $\pm 10\%$.

The magnitude of the limits is often based upon the accuracy of the measurement techniques to be used. Poor measurement techniques may justify larger limits. However, some companies maintain enterprise project management (EPM) methodologies that define the tolerances for each metric. This occurs mainly in organizations reasonably mature in project management and with some experience in metrics management. It is also possible, though uncommon, for the business case of the project to identify the critical metrics, the targets, and the tolerances.

The decision was made that each metric would have an upper and/or lower boundary established at the beginning of each project. The boundaries could change as the projects progress. If the value of the metric remained within the upper and/or lower boundaries (i.e., tolerances from the nominal value), then the metric would not appear on the crisis dashboard. Only those metrics that were above or below the tolerances or threshold limits would appear.

LXT's management was still uncertain about which metrics would indicate a crisis rather than just a potential problem. The answer was determined to be the potential damage that can occur. LXT prepared a list stating that, if any of the following can occur, then the situation would most likely be treated as a crisis:

- There is a significant threat to the outcome of the project.
- There is a significant threat to the organization, its stakeholders, and possibly the public.
- There is a significant threat to the firm's business model and strategy.
- There is a significant threat to worker health and safety.
- There is a possibility for loss of life.
- Redesigning existing systems is now necessary.
- Organizational change will be necessary.
- The firm's image or reputation will be damaged.
- There is degradation in customer satisfaction that could result in a present and/or future loss of significant revenue.

Questions

1. Is the concept of crisis dashboards workable?
2. What are some of the potential issues that LXT may face initially when using crisis dashboards?
3. Should there exist a uniform method for assigning threshold limits for the boundary boxes or should each project team be allowed to establish its own limits?
4. Can the crisis metrics be specifically related to the type of innovation such as incremental or disruptive?

(Continued)

Case Study (Continued)

5 Is it possible that information from more than one metric is needed to determine the severity of a crisis and the other metrics needed are not displayed on the dashboard because they are within the tolerance limits?

6 Are there any risks in allowing project teams to change the tolerance limits as the project progresses?

Reflections: Such as in the case study mentioned earlier, setting the tone, is basically an exercise of prioritizing. It is about highlighting, in a given culture or an organization, what is important around here, what we value, and where we need to dedicate our time, resources, and attention. Similar to how LXT management defined what will be considered a crisis, setting a tone for experiencing in an origination requires the safety in the environment previously addressed in this work. It is also a proactive exercise, just as they viewed the crisis management approach. As previously highlighted in the review of the impact of digitalization, teams could be empowered across a given portfolio with a level of transparency that makes the cascading of the strategic tone easier and more impactful.

Tip

Setting the strategic tone is critical to creating an experience-driven culture. The tone needs to be clear, widely reflected in behaviors, and continuously supported.

8.3 Value-Based Decision-Making

In our article, Kerzner and Zeitoun (2022), we tackled the criticality of ensuring that future decisions are value-based and took a portfolio view for how to enhance decisions' effectiveness.

8.3.1 Introduction

Modern-day project management has existed for almost 60 years. There have been several changes that have taken place during this time, but perhaps none as important as the acceptance and use of portfolio project management (PPM) practices.

PPM practices are based upon the type of portfolio. For simplicity's sake, portfolios can be defined as operational or strategic. With an operational PMO, the PPM team is actively involved with the individual projects, perhaps on a daily basis, and utilizes the earned value management system (EVMS) as do the project teams. On strategic projects, the PPM team interfaces mostly with stakeholders, customers, and senior management, and may require other information systems.

The type of portfolio determines the size of the portfolio and the decisions to be made. This then determines the actors that are involved in the collaboration process. The most critical business decisions reside within the strategic portfolio, and the intent of this paper is to address these strategic decisions while keeping in mind the ways of working and the experience of affected key stakeholders.

8.3.2 Experience Focused PPM

Some people seem confused as to what PPM means. PPM is NOT the same as the management of individual projects and programs. Rather, the focus is on managing the RIGHT set of projects necessary to support strategic goals and objectives as well as the expected business benefits and business value. There are generally three critical tasks for PPM:

1) To make sure that we are managing the right projects and the right number of projects in the portfolio
2) To make sure that the portfolio is composed of the right types and mix of projects
3) To make sure that all the projects are aligned to strategic goals and objectives

Most strategic portfolios do not contain all of the projects. There usually exists a prioritization process that limits the size of the portfolio. To perform these three critical tasks, the personnel assigned with the PPM responsibility must rely heavily upon their collaboration and decision-making skills.

PPM teams do not make decisions in isolation. The strategic importance of the projects within the portfolio and the decisions to be made require that the PPM team have access to all levels of the organization as well as stakeholders. The decision-making process is challenging and complex because each of the actors has their own special interests that may conflict with company interests. The ability to make meaningful and timely decisions is difficult because most companies do not have the correct decision-making tools necessary. The EVMS does not provide the critical information that most PPM teams need for strategic decisions.

8.3.3 Impact of Innovation on Decision-Making

Strategic PPM is often regarded as the collaborative process by which organizations manage innovation. Innovation drives the decisions necessary to manage the three critical PPM activities mentioned previously. All of the actors and stakeholders understand the need for innovation and as expected, may have hidden agendas where they believe that their innovation needs should be given the highest priority. The need for innovation can result in removing some projects from the portfolio and inserting new ones. This forces the PPM team to live with constantly changing information.

The PPM team must manage information from the top of the organization down to the bottom, and back up again. This also requires providing stakeholders the necessary information regarding how innovation can impact the decisions they must make now and possibly in the future.

The PPM team acts as the guardian of the critical information necessary for composition or redesign of the portfolio. The PPM team may not participate in selection of new projects or the prioritization process. The PPM team serves as the collaborative link providing the necessary information to all key actors.

One of the biggest mistakes executives make is forming a PPM team and setting high expectations without recognizing the tools that the PPM team needs to support effective decision-making. The traditional EVMS does not provide the necessary information to support decisions for idea selection and prioritization. New digitally enabled tools are needed that allow the organization to convert ideas into reality based upon some evidence and facts.

The decision-making tools must support knowledge transfer across all organizational levels that must participate in the decisions regarding portfolio content. Fortunately, developments in digital technologies are making this achievable. The digitalization tools must make sure that the organization's strategic objectives are supported by the right mix of projects. However, companies are just beginning to understand how digitalization can impact the interactions among the actors during decision-making given that the actors are at different levels of management and may have specialized interests.

8.3.4 Customer Focused Decision-Making Metrics

Strategic PMM teams require different metrics than operational PPM teams. The metrics needed are based upon the questions that the PMM team must ask themselves to provide the necessary information for the decisions the actors must make. A typical list of questions that illustrate the metrics that should be part of PMM tools include:

- Do we have any weak investments that should be canceled or replaced?
- Must any of the projects or programs be consolidated?
- Must any of the projects or programs be accelerated or decelerated?
- How well is the portfolio aligned to strategic business objectives and customers?

- Does the portfolio have to be rebalanced?
- What impact will innovation have on the composure of the portfolio?

The answer to these questions will provide information that the actors need for the decisions that they must make. Decision-making questions that actors may have include:

- Can we verify that business benefits and business value are being created and that it meets our and customers' expectations?
- What information is available to predict future performance?
- What are the strategic risks of meeting expectations and are the risks being mitigated?
- Are there any indications as to which projects may require our immediate intervention?
- How do we confirm that the portfolio is correctly aligned to our strategic goals and objectives?
- Do we need to perform resource re-optimization efforts and do we have any capacity planning restrictions impacting the size of the portfolio?

8.3.5 The Path Forward

The reflections on PPM in this paper build on the previous experience culture articles. The PPM teams of the future will be equipped with digital solutions and are expected to develop a closer understanding of the business and key customers' strategies in order to ask and address the critical set of portfolio questions.

One-size-fits-all PPM will also not exist in the future. The ingredients and building blocks around portfolio excellence will require leading with a laser focus on the experiences of the key actors in prioritizing and associated decisions. This will be affected by the type of executive sponsorship, the kind of innovation culture, and the leadership excellence that has been built in the organization. Data analytics will enhance the objectivity used by the executives and future leaders, yet they are expected to develop their strategic sensing muscles and skills.

The path forward requires a strong commitment to understanding the context of how we make our most critical strategic and investment decisions. The necessary information, strong business acumen, and understanding of the prevailing culture in the organization and its actors directly support the quality of the future portfolio decisions.

We believe the future will see continuation of managing our strategic portfolios by projects. Therefore, project managers are expected to grow their business acumen and insist on access to high-quality portfolio linkages and business information. A digitally enabled and learning-hungry next-generation leader will be able to mature our PPM practices and their true alignment with the experiences of key initiatives' stakeholders.

Reflections: As highlighted by this article, there are a few qualities to support building effective decision-making muscles. Figure 8.2 reflects the most critical ones that support the experience-driven culture. Decisions have to be value-based and the leader has to question the proposed decisions in how they map to a clear definition of value and how the success across various portfolio trade-offs will be measured. Supporting experience in this context means that the various voices, customer inputs, and stakeholders' views have been taken into the mix while adapting the decision to ensure that the ownership for the decisions has taken place. Finally, the data-enabled component ensures that there is enough considered objectivity in the decision. The trends and other analytics that could be extracted from data contribute to enhancing the confidence in the decisions taken.

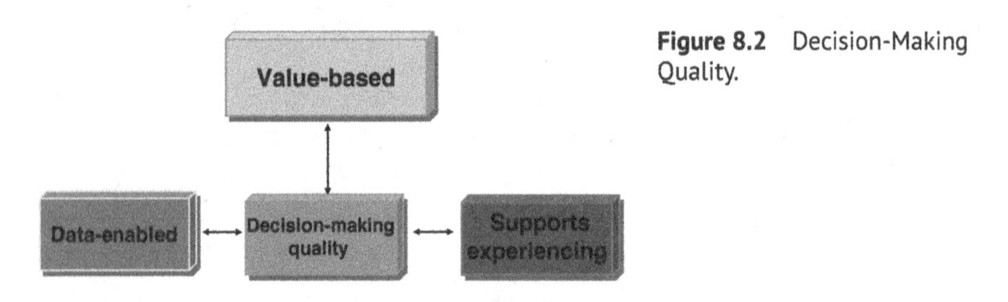

Figure 8.2 Decision-Making Quality.

> **Tip**
> Linking decisions to value is a critical aspect of the quality of these decisions. It is imperative that leaders adapt the decision-making process.

8.4 Building Momentum

Building and growing a momentum for experiencing into the future, is a critical cultural shift. As organizations build their portfolio management muscle, it is critically valuable that the portfolio mix has enough of the right number and mix of initiatives that focus on an enterprise-wide view of value. Naturally achieving a short-term set of financial ROIs from the initiatives is key, yet having a strategic view of the portfolio requires the decision makers to add components to the portfolio that have wider long-term implications.

Most critical to building this momentum, are the choices we make for how we work, the practicing of iterative principles that allow for progressive elaboration of ideas, and how the work gets done across teams and teams of teams that are tackling different parts of the portfolio. To get these principles to stick, a change

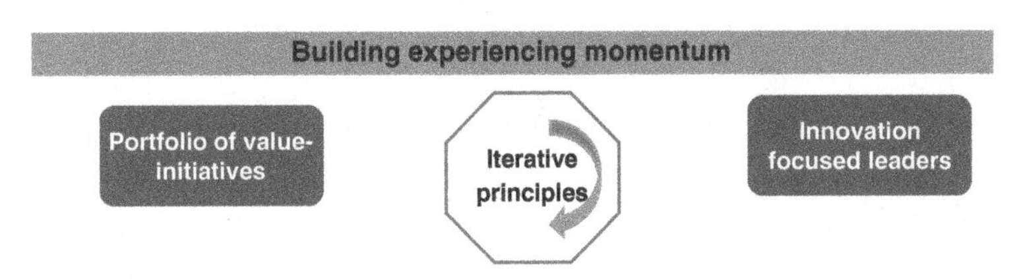

Figure 8.3 Building Momentum.

management roadmap should be put in place, and the required change champions should support how this is reflected across the business units and the various teams.

In addition, the focus and mindset shift of the leaders contribute to the likelihood of this momentum to prevail. Being focused on innovation is a critical attribute to the establishment of this momentum. This usually reflects the right risk appetite, the courageous ability to add and drop projects into the portfolio mix, and doing what is right for the overall strategic objectives (Figure 8.3).

> **Tip**
>
> Building experiencing momentum requires continual investment in the practices and mindset shifts that reward this way of working across the portfolio's projects.

References

Kerzner, H. and Zeitoun, A. (2022). The experience focused portfolio decision-making: the great project management accelerator, series article. *PM World Journal* XI (IX).

> **Review Questions**
>
> *Parentheses () are used for Multiple Choice, when one answer is correct. Brackets [] are used for Multiple Answer, when many answers are correct.*
>
> 1 What is the analogy between sports and revisiting success definition?
>   ( ) Making sure that there is a static view of success.
>   ( ) Motivating players to do whatever they like.
>   ( ) The high degree of flexibility and willingness to change tactics.
>   ( ) Everything is a sport.

(Continued)

(Continued)

2 What were some of the examples of the importance of setting a clear strategic tone? Choose all that apply.
[ ] Aligning the organization on the importance of experiencing.
[ ] Project leaders have flexibility for more changes.
[ ] Maximizing the number of projects.
[ ] Providing leading by example in support of the communicated tone.

3 What is a core principle in LXT case study that supports building effective experiences?
( ) Set fixed targets.
( ) Having various definitions for a crisis.
( ) Setting targets that have a range of tolerance to give room for variation.
( ) Ensure continual metrics reporting.

4 What is an example of a benefit of value-based decision-making?
( ) Put emphasis on using EVMS.
( ) Pushing decision-making downwards.
( ) Better support for the decisions.
( ) Higher focus on the short-term.

5 What are the three attributes that support decision-making quality? Choose all that apply.
[ ] Value-based.
[ ] Financially attractive.
[ ] Data-Enabled.
[ ] Supporting experiencing.

6 What is a key contributor to building momentum for experiencing?
( ) More projects in the portfolio.
( ) Innovation-focused leadership.
( ) Automating portfolio management.
( ) Rewarding adding more initiatives.

7 What is the meaning of progressive elaboration?
( ) Gates-based governance process.
( ) Having the ability to apply iterations for better solutions.
( ) Less reviews of decisions made.
( ) Smaller teams involved in decisions.

9

Sustaining Cultural Excellence

Leaders create experience-driven cultures. These leaders also have the responsibility of making sure that the excellence foundation built will also be sustained. Excellence is about the repeatability of certain healthy attributes in the culture, in order to mature the common practices across the organization and its teams.

Commitment to sustaining cultural excellence requires alignment to what good looks like, sensible unified practices, and ongoing supporting behaviors to continue the cascading of healthy patterns across the organization. Central to this is building an ownership culture where everyone across the organization has some level of entrepreneurial skills to support viewing the organization as one's own business. This contributes to a prevailing amount of trust and the necessary respect that is at the core of sustaining the experiencing muscles.

Key Learnings

- Understand the value of creating the jointly agreed to organizational ethos that drives connectedness across the teams.
- Learn the principles of resiliency in leading through growing uncertainty and disruptions.
- Explore the principles of building an ownership culture and how they contribute to the achievement of strong portfolio outcomes.
- Understand the mechanics of building the trust currency that is instrumental for spreading the experiencing practices across the organization.
- Develop the focus on the right supportive behaviors for supporting and sustaining excellence.

9.1 Simplicity of Powerful Culture Ethos

Martin Luther King Jr. established a powerful ethos that remained in our culture for generations. He preached nonviolent resistance, drove for equality, and wanted

Creating Experience-Driven Organizational Culture: How to Drive Transformative Change with Project and Portfolio Management, First Edition. Al Zeitoun.

to achieve justice as a core to his movement. Through repeated and consistent messages across his speeches, he exemplified focus on moral clarity and commitment to peaceful demonstrations. Ethos like Dr. King's are inspiring. He was able to inspire millions to join his movement toward racial equality and social justice.

In the space of experiencing culture, one would think of Apple Inc. as an example of an organization that has put in place a culture that puts emphasis on innovation as their ethos. One could see that the focus on excellence, user-enhanced experiences, and ultimately it would require the organization to set the features of the culture to what was previously presented in this work. Elements such as creativity, and especially simplicity, would be highly essential to align the many design, development, marketing, and other cortical teams toward sustaining one clear innovation focus for the organizational culture.

Organizations benefit from ethos that connect across the organizational culture, give people a powerful vision, and assure strong commitment to achievement of joint outcomes. For portfolio leaders to push powerful ethos across the organization, leading by example is key. Showing excellence in how actions take place and how decisions are made, is key. An instrumental part of leading by example is how an ethos gets communicated. Simplicity is a fundamental value. The worst that could happen, is that ethos gets developed and then gets lost in translation. Team members should fully understand what excellence looks like and how specifically their individual roles contribute to it.

Figure 9.1 shows a possible mix of ingredients that are necessity for building power culture ethos. Powerful ethos is timeless and can grow with the organization as it encounters external pressures, struggles with internal challenges, or has to

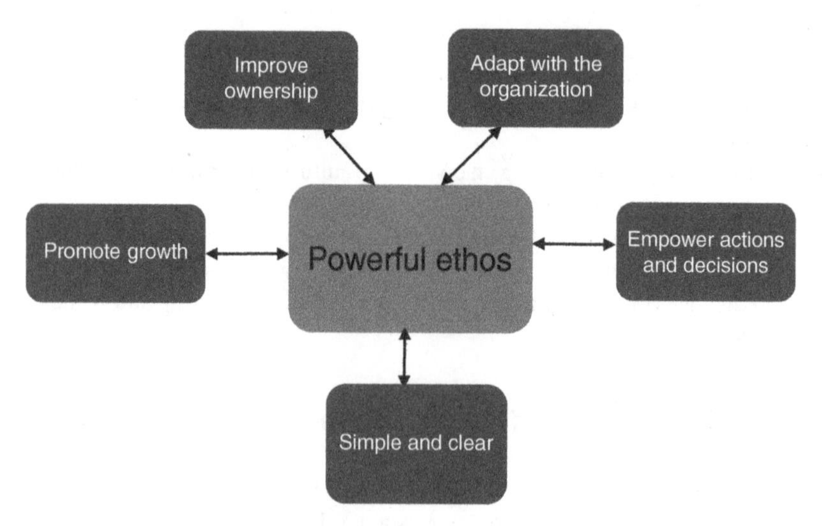

Figure 9.1 Building Powerful Ethos.

adjust to increasing customers' expectations. Critically important is how much does the ethos empowers people for action and strengthens the decision-making muscle. This is very valuable in the portfolio management process success.

As highlighted previously simplicity in this case, would mean that all are aligned fully on the meaning behind the ethos, the values linked to it, and the expectations and behaviors that support it. No silos should exist in the way of that simplicity. The organizational learning culture is also important in building effective ethos.

The value of continual growth and stretching is a fundamental criterion for building the simple and powerful ethos. Ultimately ethos is there to build an ownership culture. As will be addressed in an upcoming part of this chapter, ownership is an elevated level of accountability that will allow project and program team leaders and members to confidently deliver on complex outcomes with excellence.

Tip

In building the right supportive ethos drives connectedness. Sustaining cultural excellence is enriched by the simplicity and the empowerment an ethos can provide.

9.2 Resiliency in Leadership

Building resilience in organizations is quite similar to building an experience-driven culture on the foundational features and ingredients previously addressed in this work. Over the years, adaptability and relicense became a trademark for leaders able to drive organizations through increased chaos, unprecedented market pressures, pandemics, and other levels of uncertainty. Resilience in operations and growth requires a mindset of committing to learning and growth. It is also aligned with the psychological safety points previously addressed in choosing the right inspiring and simple ethos and how to cascade it within the teams.

In addressing one of the eight hypotheses behind this work related to achieving sustained experience, the topic of mindset and resiliency was tackled. The hypothesis was formulated as follows: "***Sustainable value creation is a mindset shift***."

The key question used to test this hypothesis was: "*What contributes the most to sustaining value creation across the organization?*"

The results in Figure 9.2 reflect the spread of the poll outcomes' scores. The mindset shift toward value was the dominating high score. It reflects a core ingredient for how resiliency could be achieved and maintained across organizations and their portfolio work. Also striking in this brief polling is the zero votes on enhancing processes. Part of the shift we are seeing toward resiliency is likely the higher dependence on values and principles to guide how work is

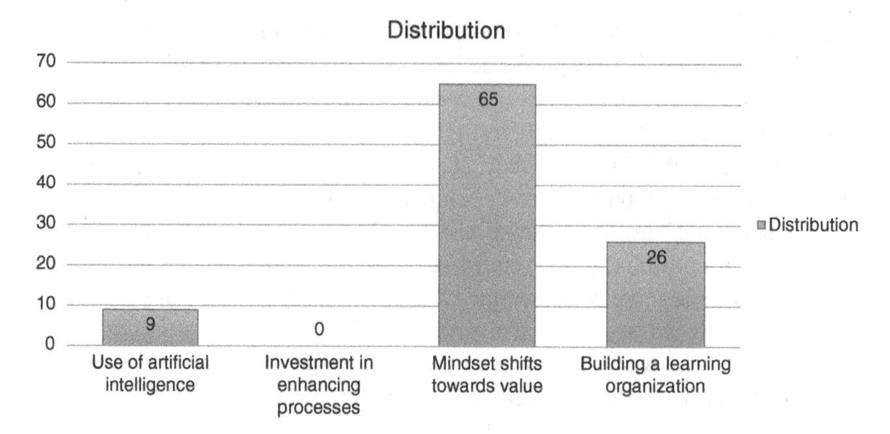

Figure 9.2 Sustaining Value Creation. *Note*: Based on LinkedIn Open Polling, April 2024.

done, more than processes unless we are tackling regulations and other industry standards.

Looking back at the movie "**Tommy Boy**" mentioned previously in this work, Tommy showed a strong model of resiliency. In that movie, as he managed to turn things around for that manufacturing organization and build his credibility as a leader, he was able to exemplify some of the resiliency-supporting ingredients. He was able to show that such a resiliency leadership quality could be built. The future of work and the increased experiencing will require change initiative leaders to exhibit passion, persistence, and patience. These are among the many of the sub-elements of designing the resilient reader.

> **Tip**
> Sustaining cultural excellence requires resilient leaders. These leaders are passionate about the organizational ethos and exemplify experiencing qualities in action.

9.3 The Ownership Culture

The strategic view organizations have directly contributes to building an ownership culture. Ownership is a more impactful word than accountability. Although both words address the commitment leaders need to have toward achieving outcomes across their teams, ownership has a sense of the experiencing and entrepreneurial dimensions that differentiate and showcase

Figure 9.3 Member of the NASA Team. Credit: geralt/Pixabay.

high-performing team quality ingredients. Portfolios or programs and projects will likely consistently succeed in the future, not because of increased governance, but due to a level of autonomy given the trust that has been built across team members.

As seen in Figure 9.3 with a janitor illustration, the most resonating ownership story could be the one from the time of NASA getting ready to execute on President JFK's mission of landing a man on the moon and returning him safely to Earth. As the President was visiting NASA to check on the progress of the work, he ran across a janitor in the hallway and asked him about what he did for NASA and the response was something like: Mr. President, I am part of the team that is putting a man on the moon."

This is a true example of a strong team, committed to one another and ultimately exhibiting a clear focus on owning joint outcomes. This is the secret sauce we need to have across portfolios of programs and projects in order to sustain excellence.

Ownership culture is a strategic priority that sponsors and executive team members should invest time and energy in achieving. As experiencing elements of cultures are being emphasized, ownership for projects' outcomes and benefits are transferred to the lowest level of the chain.

The following case study highlights a number of topics pertaining to building a culture of ownership. It would be useful to compare many of the issues raised in the case with the features and ingredients typical of creating the experiencing culture.

Case Study

The Executive Director[1]

Background

Richard Damian was delighted that his political party had won the election. As a reward for his years of support, he was appointed executive director of this government agency, replacing a person from the other political party. Damian had been with the government for more than 30 years. This would be at least a four-year appointment, and, if his party was still in power after the elections four years from now, he could be the executive director for an additional four years or more.

Damian knew how to play political gamesmanship. He avoided anything that was considered controversial and voted with his party line on all issues even if he disagreed with his party's position. He knew how to get things done behind the scenes and without exposing himself to any risks or being scrutinized by the media. But now, as executive director, he realized that things might be different. He was now exposed to the media.

The Internet Security Project

Damian's predecessor had been plagued by Internet hackers who were getting access to some of the agency's proprietary information. The media was aware of this, and his predecessor had been engulfed with bad publicity. The media kept asking what Damian was planning to do to correct the situation, and Damian kept stating that he was not ready to discuss this until he had worked out a plan with his executive staff.

1 Kerzner, 2022/John Wiley & Sons.

Damian's predecessor had tried unsuccessfully to correct the problems using the agency's internal information technology (IT) resources. Unfortunately, the internal resources had limited IT security knowledge. Budgetary cuts made it impossible to hire additional IT resources. Government hiring and firing practices also made it difficult to remove some poor performers and replace them with other workers trained in IT security practices. Damian's predecessor also tried to get support from other government agencies, but they had their own political agendas and could not or would not provide the needed support.

The project had to be outsourced. Damian instructed one of his direct reports to assign someone as the project manager and begin by soliciting bids from at least three vendors. Since time was critical because of the pressure imposed by the media, Damian recommended that the quotes be obtained informally because of the time-consuming, rigid policies and procedures that must be followed for traditional government contracting.

Assigning the Project Manager

The person assigned as the project manager reported several levels below Damian. The project manager had been with the agency for less than two years and had a degree in information systems. In addition, the project manager, and the rest of the agency's IT group, had very little knowledge about IT security. The agency would have to rely on the expertise of contractors.

Damian believed that rank has its privilege. As such, he decided that he should not have to interface directly with people who report low on the organizational chart. He instructed one of his direct reports to keep him informed about the status of the project.

A month later, Damian was informed that there were three qualified bidders. The bids ranged from US$ 1.5 million to US$ 1.75 million with a time frame of approximately three months. All three bidders said that their bid was just a rough estimate and that a final bid could not be made without a clear examination of the agency's existing hardware and software. Furthermore, all three bidders stated that they wanted a cost-reimbursable contract rather than a firm-fixed-price contract.

Damian was in a hurry to get the contract started. He instructed the procurement people to issue a cost-reimbursable contract to one of the vendors immediately. This required violation of traditional procurement policies and procedures, but Damian felt that this was an extraordinary situation that needed resolution quickly. Two weeks later, the contract was signed and the project had a go-ahead date of March 1.

(Continued)

Case Study (Continued)

The Work Begins

As soon as the work began, Damian held a news conference and announced that the security system was being modified and that all protocols for the new system would be operational within 90 days, the duration of the contract. Even though he had very limited knowledge about how the system would work, he still made promises concerning the system's capabilities. The media seemed somewhat skeptical about how quickly the changes would be made and Damian's promises and began asking questions. Damian knew how to play the political game. He certainly did not have enough information to answer the questions that might be forthcoming. He declined to answer any of the media's questions, stating that all questions would be answered at a future new conference.

Over the next month, the media kept asking why Damian's agency was not providing any information on the status of the new security system. It was rumored that the project would be coming in late and over budget. Damian was several layers of management removed from where the work was taking place and knew very little about the progress of the project. Information on the status of the project, especially any bad news, was being filtered out as the information flowed up the organizational hierarchy to the point where it appeared that there were no issues. Unfortunately, that was not the case.

Damian learned that the project would be at least one month late and possibly over budget by US$ 500,000. He called a news conference and informed the media about the schedule slippage and cost overrun. Trying to protect his image, Damian stated that he was never informed about the risks on the project. Furthermore, he said that he was forming a committee headed by one of his direct reports to get to the bottom of the problem. Once again, Damian refused to answer any questions posed by the media.

The Problems Mount

Damian asked his direct report for a briefing on what was being done to correct the situation. The information provided by the contractor stated that his agency's hardware and existing software were outdated and needed to be replaced. The software and changes needed to enhance computer security could not run on the existing hardware. The contractor was now asking for an additional US$ 4 million to update all of the hardware and software. The agency's IT personnel were all in favor of the upgrade. The entire changeover would add six months to the length of the schedule.

The US$ 1.75 million project was now at US$ 6 million and possibly increasing. Damian was convinced that the media would attack his credibility because of the security issues that still existed. Someone had to be blamed so that the "heat" would not be placed on Damian. His first thought was to blame the project manager, but everyone would know that this was not the case.

Playing the political game, Damian called another news conference and blamed his predecessor for all of the existing issues. He stated that these problems should have been addressed years ago and that his predecessor, who belonged to a different political party, failed to take the necessary steps to correct the situation. Furthermore, Damian asserted that one of his direct reports was now the sponsor for the entire project and that Damian would receive weekly progress briefings.

Damian played the political game to the best of his ability. Not only did he blame the other political party and his predecessor for all of the problems, but he insulated himself from further disasters by stating that one of his direct reports was now the sponsor. Damian now believed that he would be free from further criticism.

The Situation Worsens

Damian's agency did not have authorization for purchasing computer hardware and software without getting permission from the government's centralized procurement group. The hardware and software recommendations from Damian's contractor were not on the government's approved hardware and software list. The contractor either had to select hardware and software from the approved list or apply for add-ons to the list. Applying for add-ons to the approved list could take as much as three to six months, thus lengthening the existing contract.

The contractor reviewed the list and recommended purchasing hardware and software that would increase the project's cost by US$ 5 million rather than US$ 4 million. The length of the project was now estimated to be 1.5 years rather than three months, assuming, of course, that the hardware could be received within a reasonable amount of time. Trying to get additional hardware and software added to the approved list could have possibly saved US$ 1 million but might have lengthened the project to two years.

The media became aware of the situation and began attacking Damian's credibility as an executive director. Once again Damian had to play the political game. He called a news conference and stated that, although time and cost were both constraints, time was prioritized as being more important than

(Continued)

Case Study (Continued)

cost. Therefore, he accepted the cost overrun of US$ 5 million. Furthermore, Damian stated that he would now be the sponsor for this project, although executive directors normally do not sponsor projects of this size.

Damian once again blamed his predecessor for all of the problems, stating that this problem could have been done years ago at a lower cost. He also left the media with the impression that his direct report, who was previously the sponsor, was being reassigned to another position. That move made it appear that his direct report was partially at fault.

The Project Approaches Completion

The original project, with a budget of US$ 1.75 million and a duration of three months, was completed at the end of 2.5 years and at a cost of US$ 9 million. Software bugs were found during testing that required overtime, software changes, and the purchase of some additional hardware.

Once again Damian played the political game. He informed the media that the project was completed and that security was now in place. Furthermore, he stated that he personally was in charge of the project all the way and took full credit for its successful completion.

Questions

1. Could the solution to the security problem have been managed internally, or was it necessary to outsource the work?

2. Because of the sensitivity of the problem, was it acceptable to get quotes from vendors on an informal basis, or should everything have been done through the formal channels for procurement?

3. Why did Damian want someone quite low in the organizational hierarchy to function as the project manager?

4. Why did Damian want one of his direct reports to keep him informed about the status of the project rather than hearing it directly from the project manager?

5. Why did all three bidders want a cost-reimbursable contract rather than a fixed-price contract?

6 Why was Damian reluctant to address the media and answer their questions once the project started?

7 What was the driving force behind Damian's actions throughout the case?

Reflections: Not only does the executive director exhibit all the signs of leadership weaknesses, but he also illustrated terrible examples of what is typically needed to build an ownership culture. Starting such a high-visibility project with a project manager who was not properly authorized, or had any proper coverage of a proper project sponsor, was already a poor demonstration of building ownership. Later, having a direct report become the sponsor, not paying attention to building the critical transparency, and allowing the many layers of bureaucracy in the organization to filter risks and key information, all contributed to destring trust and breaking the foundation of creating ownership.

Even when finally taking the role of the sponsor, the executive director continued to blame others and never took true ownership of the growing list of issues, extreme deviations in cost and schedule, and the lack of a proper vision that would have otherwise connected the stakeholders of this critical initiative.

Tip

Ownership cultures are critical for experiencing. They are built on a foundation of trust, transparency, and delegation of authority, and are supported with proper championing.

9.4 Trust Foundation Building

Just like with the classical views of Patrick Lencioni highlighted in his work, "The Five Dysfunctions of Teams and How to Overcome Them," having a focus on building a culture of ownership would have meant that teams have done their homework in creating the open environment where everyone could support the trust foundation with the ability to have healthy conflicts openly within a safe dialogue setting.

The Center for Creative Leadership indicates: "75% of careers are derailed for reasons related to emotional competencies, including inability to handle interpersonal problems; unsatisfactory team leadership during times of difficulty or conflict; or inability to adapt to change or elicit trust."

Figure 9.4 The Trust Currency. Credit: Breedstock/Pixabay.

Teams in the future will sustain their cultural excellence by focusing on the currency of trust. As highlighted in Figure 9.4 the trust currency of the future will also include a digital element. Teams will not only have to invest in the human-to-human trust building and in designing joint commitment, but they will also have to ensure that we enable the digital trust in the data, the safeguarding of the organizational intellectual capital, adhering to legal requirements, and addressing the associated cyber security concerns.

To build a foundation of trust, management has to set the strategic tone in the culture and enable the creation of an open and transparent environment that is also learning-based, as previously addressed in this work as one of the features of experiencing culture. Project and program teams have to continuously challenge their abilities to handle tough challenges, ensure positive open discussions around differences of opinion, and continue to be inclusive and inspiring in their leadership style.

Tip
Building a trust foundation protects the ownership culture. This also contributes to enhancing the decision-making excellence needed for effective portfolio execution.

9.5 Sustaining Supporting Behaviors

One of the most inspiring places to work as a project manager could be Imagineering. As covered in the following case study, sustaining the experience-driven culture requires creativity, ideas, and behaviors that leaders should embody to leave a lasting expression from projects' outcomes, like one would encounter in Disney theme parks. The case highlights how to sustain the many experiencing features previously addressed in this work.

Case Study

Disney: Imagineering Project Management[2]

Introduction

Not all project managers are happy with their jobs, and they often believe that changing industries might help. Some want to manage "the world's greatest construction projects" while others want to design the next-generation cell phone or mobile device. However, the project managers who probably are the happiest are the Imagineering project managers who work for the Walt Disney Company, even though they probably could earn higher salaries elsewhere on projects that have profit and loss statements. Three Imagineering project managers—John Hench, Claude Coats, and Martin Sklar retired with a combined 172 years of Imagineering project management work experience with the Walt Disney Company. But how many project managers in other industries truly understand what skills are needed to be successful as an Imagineering project manager? Is it possible that many Imagineering project management skills are applicable to other industries and we do not recognize it?

The Project Management Body of Knowledge (PMBOK®) Guide is, as the name implies, just a guide. Each company may have unique or specialized

(Continued)

2 Kerzner, 2022/John Wiley & Sons.

Case Study (Continued)

skills needed for the projects it undertakes above and beyond what is included in the PMBOK® Guide. Even though the principles of the Guide apply to Disney's theme park projects, other skills are needed that are significantly different from much of the material taught in traditional project management courses. Perhaps the most common skills among all Imagineering project managers are brainstorming, problem-solving, decision-making, and thinking in three rather than two dimensions. While many of these skills are not taught in depth in traditional project management programs, they may very well be necessities for all project managers. Yet most of us may not recognize this fact.

Walt Disney Imagineering

Walt Disney Imagineering (also known as WDI or simply Imagineering) is the design and development arm of the Walt Disney Company, responsible for the creation and construction of Disney theme parks worldwide. Founded by Walt Disney to oversee the production of Disneyland Park, the company was originally known as WED Enterprises, from the initials meaning "Walter Elias Disney," the company founder's full name.[3]

The term "Imagineering" was introduced in the 1940s by Alcoa to describe its blending of imagination and engineering, and used by Union Carbide in an in-house magazine in 1957, with an article by Richard F. Sailer called "BRAIN-STORMING IS IMAGINation enginEERING." Disney filed for a copyright for the term in 1967, claiming first use of the term in 1962.

Imagineering is responsible for designing and building Disney theme parks, resorts, cruise ships, and other entertainment venues at all levels of project development. Imagineers possess a broad range of skills and talents, and thus over 140 different job titles fall under the banner of Imagineering, including illustrators, architects, engineers, lighting designers, show writers, graphic designers, and many more.[4] It could be argued that all Imagineers are project managers and all project managers at WDI are Imagineers. Most Imagineers work from the company's headquarters in Glendale, California, but are often

3 Alex Wright, Imagineers: The Imagineering Field Guide to Magic Kingdom at Walt Disney World (New York: Disney Editions, 2005).
4 Ibid.

deployed to satellite branches within the theme parks for long periods of time.

Parts of this case study have been adapted from Wikipedia contributors, "Walt Disney Imagineering," Wikipedia, The Free Encyclopedia, https://en.wikipedia.org/w/index.php?title=Walt_Disney_Imagineering&oldid=758012775

Project Deliverables

> All I want you to think about is when people walk through or have access to anything you design, I want them, when they leave, to have smiles on their faces. Just remember that. It's all I ask of you as a designer.
>
> —Walt Disney

Unlike traditional projects where the outcome of a project is a hardware or software deliverable, Imagineering project outcomes for theme park attractions are visual stories. The entire deliverable is designed to operate in a controlled environment where every component has a specific meaning and contributes to a part of telling a story. It is visual storytelling. Unlike traditional movies or books that are two-dimensional, theme parks and the accompanying characters come to life in three dimensions. Most project managers do not see themselves as storytellers. The intent of a theme park attraction is to remove people from reality once they enter the attraction and make them believe that they are living out a story and possibly interacting with their favorite characters. Theme park visitors of all ages are made to feel that they are participants in the story rather than just observers.

Some theme parks are composed of rides that appeal to just one of your senses; Disney's attractions, in contrast, appeal to several senses, thus leaving a greater impact when people exit the attraction. "People must learn how to see, hear, smell, touch and taste in new ways."[5] Everything is designed to give people an experience. In the ideal situation, people are made to believe that they are part of the story. When new attractions are launched, Imagineers pay attention to guests' faces as they come off of a ride. This is important for continuous improvement efforts.

(Continued)

5 John Hench with Peggy Van Pelt, Designing Disney: Imagineering and the Art of the Show (New York: Disney Editions, 2008), p. 2.

Case Study (Continued)

The Importance of Constraints

Most project management courses emphasize that there are three constraints on projects, namely time, cost, and scope. Although these constraints exist for Imagineering projects as well, there are three other theme park constraints that are often considered more important than time, cost, and scope. The additional constraints are safety, quality, and aesthetic value.

Safety, quality, and aesthetic value are all interrelated constraints. Disney will never sacrifice safety. It is first and foremost the primary constraint. All attractions operate every few minutes 365 days each year and must satisfy the strictest of building codes. Some rides require special effects, such as fire, smoke, steam, and water. All of this is accomplished with safety in mind. Special effects include fire that actually does not burn, simulated fog that people can breathe safely, and explosions that do not destroy anything. Another special effect is the appearance of bubbling molten lava that is actually cool to the touch.

Reliability and maintainability are important quality attributes for all project managers but are of critical importance for the Imagineers. In addition to fire, smoke, stream, and water, there are a significant number of moving parts in each attraction. Reliability considers how long something will perform without requiring maintenance. Maintainability concerns how quickly repairs can be made. Attractions are designed with consideration given to component malfunctions and ways to minimize the downtime. Some people may have planned their entire vacation around the desire to see specific attractions, and if these attractions are down for repairs for a lengthy time, park guests will be unhappy.

Brainstorming

With traditional projects, brainstorming may be measured in hours or days. Members of the brainstorming group are few in number and may include marketing for the purpose of identifying the need for a new product or enhancement to existing product and technical personnel to state how long it takes and the approximate cost. Quite often, traditional project managers may not be assigned and brought on board until after the project has been approved, added into the queue, and after the statement of work (SOW) is well-defined. At Disney's Imagineering organization, brainstorming may be measured in years and a multitude of Imagineering personnel will participate, including the project managers.

Attractions at most traditional amusement parks are designed by engineers and architects. Imagineering brainstorming at Disney is done by storytellers who must visualize their ideas in both two and three dimensions. Brainstorming could very well be the most critical skill for an Imagineer. It requires that Imagineers put themselves in the guests' shoes and think like children as well as adults in order to see what the visitors will see. Those who design an attraction must know the primary audience.

Brainstorming can be structured or unstructured. Structured brainstorming could entail thinking up an attraction based on a newly released animated or nonanimated Disney movie. Unstructured brainstorming is usually referred to as "blue-sky" brainstorming. Several sessions may be required to come up with the best idea because people need time to brainstorm. Effective brainstorming mandates that people be open-minded to all ideas. And even if everyone agrees on the idea, Imagineers always ask, "Can we make it even better?" Unlike traditional brainstorming, it may take years before an idea comes to fruition at the Imagineering Division.

Imagineering brainstorming must focus on a controlled themed environment where every component is part of telling the story. Critical questions must be addressed and answered as part of Imagineering brainstorming:

- How much space will I have for the attraction?
- How much time will the guests need to feel the experience?
- Will the attraction be seen on foot or using people movers?
- What colors should we use?
- What music should we use?
- What special effects and/or illusions must be in place?
- Does technology exist for the attraction, or must new technology be created?
- What landscaping and architecture will be required?
- What other attractions precede this attraction or follow it?

Before brainstorming is completed, the team must consider the cost. Regardless of the technology, can we afford to build it? This question must be addressed during structured and blue-sky brainstorming sessions.

Guiding Principles

> If I could pick any job here, I'd move my office to the Imagineering building and immerse myself in all that lunacy and freethinking.
>
> —Michael D. Eisner, former CEO, Walt Disney

(Continued)

Case Study (Continued)

When developing new concepts and improving existing attractions, Imagineers are governed by a few key principles. Often new concepts and improvements are created to fulfill specific needs and to make the impossible appear possible. Many ingenious solutions to problems are Imagineered in this way, such as the ride vehicle of the attraction Soarin' Over California. The Imagineers knew they wanted guests to experience the sensation of flight but weren't sure how to accomplish the task of loading the people onto a ride vehicle in an efficient manner where everyone had an optimal viewing position. One day an Imagineer found an Erector set in his attic and was able to envision and design a ride vehicle that would effectively simulate hang gliding.[6]

Imagineers are also known for returning to ideas for attractions and shows that, for whatever reason, never came to fruition. It could be years later when they revisit the ideas. These ideas are often reworked and appear in a different form like the Museum of the Weird, a proposed walk-through wax museum that eventually became the Haunted Mansion.[7]

Finally, there is the principle of "blue-sky speculation," a process where Imagineers generate ideas with no limitations. The custom at Imagineering has been to start the creative process with what is referred to as "eyewash"—the boldest, wildest, best idea a person can come up with, presented in absolutely convincing detail. Many Imagineers consider this to be the true beginning of the design process and operate under the notion that if it can be dreamed of, it can be built.[8] Disney believes that everyone can brainstorm and that everyone wants to contribute to the brainstorming process. No ideas are bad ideas. Effective brainstorming sessions neither evaluate nor criticize ideas. They are recorded and may be revisited years later.

Imagineers are always seeking to improve on their work, what Walt Disney called "plussing." He firmly believed that "Disneyland will never be completed as long as there's imagination left in the world," meaning there is always room for innovation and improvement.[9] Ideas and eventually future attractions can also come from the animated films produced by the Walt Disney Company or other film studios.

6 George Scribner and Jerry Rees (directors), Disneyland: Secrets, Stories, and Magic (DVD). Walt Disney Video, 2007.
7 Ibid.
8 Karal Ann Marling, Designing Disney's Theme Parks: The Architecture of Reassurance (New York: Flammarion, 1997).
9 Scribner and Rees, Disneyland.

The brainstorming subsides when the basic idea is defined, understood, and agreed upon by all group members. It belongs to all of us, keeping strong a rich heritage left to us by Walt Disney. Teamwork is truly the heart of Imagineering. In that spirit though, Imagineering is a diverse collection of architects, engineers, artists, support staff members, writers, researchers, custodians, schedulers, estimators, machinists, financiers, model-makers, landscape designers, special effects and lighting designers, sound technicians, producers, carpenters, accountants, and filmmakers, we all have the honor of sharing the same unique title.

Here, you will find only Imagineers.[10]

Imagineering Innovations

Over the years, WDI has been granted over 115 patents in areas such as ride systems, special effects, interactive technology, live entertainment, fiber optics, and advanced audio systems.[11] WDI is responsible for technological advances such, as the Circle-Vision 360° film technique and the FastPass virtual queuing system.

Imagineering must find a way to blend technology with the story. Imagineering is perhaps best known for its development of Audio-Animatronics, a form of robotics created for use in shows and attractions in the theme parks that allowed Disney to animate things in three dimensions instead of just two dimensions. The idea sprang from Disney's fascination with a mechanical bird he purchased in New Orleans, which eventually led to the development of the attraction the Enchanted Tiki Room. The Tiki Room, which debuted in 1963 and featured singing Audio-Animatronic birds, was the first to use such technology. The 1964 World's Fair featured an Audio-Animatronic figure of Abraham Lincoln that actually stood up and delivered part of the Gettysburg Address (which incidentally had just passed its centennial at the time) for the "Great Moments with Mr. Lincoln" figure exhibit, the first human Audio-Animatronic.[12]

Today, Audio-Animatronics are featured prominently in many popular Disney attractions, including Pirates of the Caribbean, the Haunted Mansion, the Hall of Presidents, Country Bear Jamboree, Star Tours: The Adventures Continue, and Muppet*Vision 3D. Guests also have the opportunity to interact

(Continued)

10 Disney Book Group, Walt Disney Imagineering (New York: Disney Editions, 1996), p. 21.
11 Walt Disney Imagineering website, www.imagineeringdisney.com.
12 Ibid.

Case Study (Continued)

with some Audio-Animatronic characters, such as Lucky the Dinosaur, WALL-E, and Remy from Ratatouille. The next wave of Audio-Animatronic development focuses on completely independent figures, or "autonomatronics." Otto, the first autonomatronic figure, is capable of seeing, hearing, sensing a person's presence, having a conversation, and even sensing and reacting to guests' emotions.

Storyboarding

Most traditional project managers may be unfamiliar with the use of storyboarding as applied to projects. At Disney Imagineering, it is an essential part of the project. Ideas at Imagineering begin as a two-dimensional vision drafted on a piece of white paper. Storyboards, which are graphic organizers in the form of illustrations or images displayed in sequence for the purpose of pre-visualizing the relationship between time and space in the attraction, assist the Imagineers in seeing the entire attraction. Storyboards also are used in motion pictures, animation, motion graphics, and interactive media. They provide a visual layout of events as they are to be seen by the guests. The storyboarding process, in the form it is known today, was developed at Walt Disney Productions during the early 1930s, after several years of similar processes being in use at Walt Disney and other animation studios.

A storyboard is essentially a large comic of the attraction produced beforehand to help the Imagineers visualize the scenes and find potential problems before they occur. Storyboards also help estimate the cost of the overall attraction and save development time. Storyboards can be used to identify where changes to the music are needed to fit the mood of the scene. Often storyboards include arrows or instructions that indicate movement. When animation and special effects are part of the attraction, the storyboarding stage may be followed by simplified mock-ups called "animatics" to give a better idea of how the scene will look and feel with motion and timing. At its simplest, an animatic is a series of still images edited together and displayed in sequence with a rough dialogue and/or rough soundtrack added to the sequence of still images (usually taken from a storyboard) to test whether the sound and images are working together effectively.

The storyboarding process can be very time-consuming and intricate. Today, storyboarding software is available to speed up the process.

Mock-Ups

Once brainstorming has been completed, mock-ups of the idea are created. Mock-ups are common to some other industries, such as construction. Simple mock-ups can be made from paper, cardboard, Styrofoam, plywood, or metal.

The modelmaker is the first Imagineer to make a concept real. The art of bringing a two-dimensional design into three dimensions is one of the most important and valued steps in the Imagineering process. Models enable the Imagineer to visualize, in miniature, the physical layout and dimensions of a concept, and the relationships of show sets or buildings as they will appear.

As the project evolves, so too do the models that represent it. Once the project team is satisfied with the arrangements portrayed on massing models, small-scale detailed-oriented study models are begun. This reflects the architectural styles and colors for the project.

Creating a larger overall model, based upon detailed architectural and engineering drawings, is the last step in the model-building process. This show model is the exact replica of the project as it will be built, featuring the tiniest of details, including building exteriors, landscape, color schemes, the complete ride layout, vehicles, show sets, props, figures, and suggested lighting and graphics.[13]

Computer models of the complete attraction, including the actual ride, are next. They are computer-generated so that the Imagineers can see what the final product looks like from various positions without actually having to build a full-scale model. Computer models, similar to CAD/CAM modeling, can show in three dimensions the layout of all of the necessary electrical, plumbing, HVAC, special effects, and other equipment.

Aesthetics

Imagineers view the aesthetic value of an attraction in a controlled theme environment as a constraint. This aesthetic constraint is more of a passion for perfection than the normal constraints that most project managers are familiar with.[14]

Aesthetics are the design elements that identify the character and the overall theme and control the environment and atmosphere of each setting.

(Continued)

13 Disney Book Group, Walt Disney Imagineering: A Behind the Scenes Look at Making the Magic Real, p. 72.

14 Some people argue that the aesthetics focuses more on creating a controlled environment than on reality, thus controlling your imagination.

Case Study (Continued)

This includes color, landscaping, trees, colorful flowers, architecture, music, and special effects. Music must support the mood of the ride. The shape of the rocks used in the landscape is also important. Pointed or sharp rocks may indicate danger whereas rounded or smooth rocks may represent safety. Everything in the attraction is there for the purpose of reinforcing a story. Imagineers go to minute levels of detail for everything needed to support the story without overwhelming the viewers with too many details. Details that are contradictory can leave the visitors confused about the meaning of the story.

A major contributor to the aesthetics of the attraction is the special effects. Special effects are created by "Illusioneering," which is a subset of Imagineering.

Special effects can come in various forms. Typical projected special effects can include:

- Steam, smoke clouds, drifting fog, swirling effects
- Erupting volcano, flowing lava
- Lightning flashes and strikes, sparks
- Water ripple, reflection, waterfall, flows
- Rotating and tumbling images
- Flying, falling, rising, moving images
- Moving images with animated sections
- Kaleidoscopic projections
- Liquid projections, bubbles, waves
- Aurora borealis, Lumia, abstract light effects
- Twinkling stars (when fiber optics cannot be used, such as on rear-projection screen)
- Spinning galaxies in perspective, comets, rotating space stations, pulsars, meteor showers, shooting stars, and any astronomical phenomena
- Fire, torches, forest fire
- Expanding rings
- Ghosts, distorted images
- Explosions, flashes[15]

15 See "Bill Novey and the Business of Theme Park Special Effects," http://blooloop.com/feature/disney-imagineering-bill-novey-and-the-business-of-theme-park-special-effects-2/. The paper provides an excellent summary of various special effects used by Illusioneers. In addition to the projected special effects, the paper also describes laser effects, holographic images, floating images, mirror gags, gas discharge effects, and fiber optics.

Perhaps the most important contributor to the aesthetic value of an attraction is color. Traditional project managers rely on sales or marketing personnel to select the colors for a deliverable. At Imagineering, it is done by the Imagineers. Color is a form of communication. Even the colors of the flowers and the landscaping are critical. People feel emotions from certain colors, either consciously or subconsciously. Imagineers treat color as a language. Some colors catch the eye quickly, and we focus our attention on them. "We must ask not only how colors work together, but how they make the viewer feel in a given situation. It is the Imagineer's job to understand how colors work together visually and why they can make guests feel better."[16]

"White represents cleanliness and purity, and in many European and North American cultures is the color most associated with weddings, and with religious ceremonies such as christenings. Silver white suggests joy, pleasure, and delight. In architecture and interior design, white can be monotonous if used over large areas."[17] "We have created an entire color vocabulary at Imagineering, which includes colors and patterns we have found that stir basic human instincts – including that of survival."[18]

Aesthetics also impacts the outfits and full-body costumes of the cast members who are part of the attraction. The outfits that the cast members wear must support the attraction. Unlike animation, where there are no physical limitations to a character's identity or mobility, people may have restricted motion once in the costume. Care must be taken that the colors used in the full-body costumes maintain the character's identity without conflicting with the background colors used in the attraction. Even the colors in the restrooms must fit the themed environment. Imagineers also try to address queue design by trying to make it a pleasant experience. As people wait in line to see an attraction, aesthetics can introduce them to the theme of the attraction. The aesthetics must also consider the time it takes people to go from attraction to attraction as well as what precedes this attraction and what follows it. "For transition to be smooth, there must be a blending of themed foliage, color, sound, music, and architecture. Even the soles of your feet feel a change in the paving explicitly and tell you something new is on the horizon."[19]

(Continued)

16 Hench with Van Pelt, Designing Disney, p. 104.
17 Ibid., p. 135.
18 Disney Book Group, Walt Disney Imagineering, p. 94.
19 Ibid., p. 90.

Case Study (Continued)

The Art of the Show

Over the years, Imagineering has conceived a whole range of retail stores, galleries, and hotels that are designed to be experienced and to create and sustain a very specific mood. For example, the mood of Disney's Contemporary Resort could be called "the hello futuristic optimism," and it is readily apparent, given the resort's A-frame structure, futuristic building techniques, modern décor, and the monorail gliding quietly through the lobby every few minutes. Together, these details combine to tell the story of the hotel.[20]

Imagineering is, first and foremost, a form of storytelling, and visiting a Disney theme park should feel like entering a show. Extensive theming, atmosphere, and attention to detail are the hallmarks of the Disney experience. The mood is distinct and identifiable, the story made clear by details and props. Pirates of the Caribbean evokes a "rollicking buccaneer adventure," according to Imagineering Legend John Hench,[21] whereas the Disney Cruise Line's ships create an elegant seafaring atmosphere. Even the shops and restaurants within the theme parks tell stories. Every detail is carefully considered, from the menus to the names of the dishes to the cast members' costumes.[22] Disney parks are meant to be experienced through all senses, for example, as guests walk down Main Street, U.S.A., they are likely to smell freshly baked cookies, a small detail that enhances the story of small-town America at the turn of the nineteenth century.

The story of Disney theme parks is often told visually, and the Imagineers design the guest experience in what they call "The Art of the Show." John Hench was fond of comparing theme park design to moviemaking and often used filmmaking techniques, such as forced perspective, in the Disney parks.[23] Forced perspective is a design technique in which the designer plays with the scale of an object in order to affect the viewer's perception of the object's size. One of the most dramatic examples of forced perspective in the Disney parks is Cinderella's Castle. The scale of architectural elements is much smaller in the upper reaches of the castle compared to the foundation, making it seem significantly taller than its actual height of 189 feet.[24]

---

20 Marling, Designing Disney's Theme Parks.
21 Hench and Van Pelt, Designing Disney, p. 56.
22 Hench and Van Pelt, Designing Disney.
23 Ibid., p. 74.
24 Wright, Imagineers.

The Power of Acknowledgment

Project managers like to be told that they have done a good job. It is a motivational force encouraging them to continue performing well. However, acknowledgment does not have to come with words; it can come from results. At Disney's Imagineering Division, the fact that more than 132,500,000 visitors passed through the gates of the 11 Disney theme parks in 2013 is probably the greatest form of acknowledgment. The Walt Disney Company does acknowledge some Imagineers in other ways. Disney established a society entitled "Imagineering Legends." Three of their most prominent Imagineering Legends are John Hench (65 years with Disney), Claude Coats (54 years with Disney), and Martin Sklar (53 years with Disney). The contributions of these three Imagineers appear throughout the Disney theme park attractions worldwide. The goal of all Imagineers at Disney may very well be the acknowledgment of becoming an Imagineering Legend.

The Need for Additional Skills

All projects have special characteristics that may mandate a unique set of project management skills above and beyond what we teach using the PMBOK® Guide. Some of the additional skills that Imagineers may need are summarized next.

- The ability to envision a story
- The ability to brainstorm
- The ability to create a storyboard and build mock-ups in various stages of detail
- A willingness to work with a multitude of disciplines in a team environment
- An understanding of theme park design requirements
- Recognizing that the customers and stakeholders range from toddlers to senior citizens
- An ability to envision the attraction through the eyes and shoes of the guests
- An understanding of the importance of safety, quality, and aesthetic value as additional competing constraints
- A passion for aesthetic details
- An understanding of the importance of colors and the relationship between colors and emotions
- An understanding of how music, animatronics, architecture, and landscaping must support the story

(Continued)

Case Study (Continued)

Obviously, this list is not all-inclusive, but it does show that not everyone can be an Imagineer for Disney. These skills also apply to many of the projects that most project managers are struggling with. Learning and applying these skills could very well make all of us better project managers.

Questions

1. Why do most project managers not recognize that they either need or can use the skills required to perform as an Imagineering project manager?
2. What is the fundamental difference between a ride and an attraction?
3. What are some of the differences between traditional brainstorming and Imagineering brainstorming?
4. How many project constraints are there on a traditional theme park attraction?
5. How would you prioritize the constraints?
6. Why is it necessary to consider cost before the Imagineering brainstorming sessions are completed?
7. What is Audio-Animatronics?
8. What is storyboarding, and how is it used on Disney projects?
9. What is meant by "project aesthetics," and how might it apply to projects other than at Disney?

Tip

Sustaining the right behaviors for cultural excellence requires open and creative storytelling. Just like in Imagineering, a great question is: "Can we make it even better?".

Review Questions

Parentheses ( ) are used for Multiple Choice, when one answer is correct. Brackets [ ] are used for Multiple Answer, when many answers are correct.

1 What is central to the ability of sustaining cultural excellence?
( ) Making sure that culture is the responsibility of the team.
( ) Motivating team members with higher rewards.
( ) Building an ownership culture.
( ) Controlled risk appetite.

2 What are some of the advantages of setting up a simple powerful ethos? Choose all that apply.
[ ] Promoting learning and growth.
[ ] Should not change across the organizational journey.
[ ] Empowering decision-making.
[ ] Improving ownership.

3 What is a possible supporting principle for enhancing leadership resiliency?
( ) Set stretching metrics.
( ) Increasing the investment in more processes.
( ) Mindset shifts toward value.
( ) Ensure full automation of mundane processes.

4 What is an example of ownership behavior of leaders?
( ) Put emphasis on the political image.
( ) Questioning every move by the stakeholders.
( ) Stepping in and taking responsibility for a crisis.
( ) Ensuring that there is a point of blame.

5 What could be other example of project constraints beyond the classical three constraints? Choose all that apply.
[ ] Use of Lava.
[ ] Number of Visitors.
[ ] Safety.
[ ] Aesthetic value.

(Continued)

(Continued)

6 What is a key contributor to building the trust currency?
( ) More time in situation analysis.
( ) Being able to have tough conversations with opposing views.
( ) Automating the trust building.
( ) Rewarding vertical work in projects.

7 What is a key skill that Imagineers have that other project managers could benefit from?
( ) Visualizing in two dimensions.
( ) Storytelling.
( ) Use of multiple metrics.
( ) Shorter brainstorming cycles.

10

Enterprise Portfolio Management Muscles

Organizations excel on purpose. The path to excellence starts with a clear strategy that connects to the organizational ethos and that drives the right actions and decisions across teams. Portfolio management muscles are the glue that could connect the dots across this ecosystem. These muscles are built on a lifecycle view of the processes and principles that support managing the portfolio. They require integrated leadership that is capable to execute along this lifecycle. It has to be enhanced with data that enables the movement of resources and investments across the portfolio and allows executives to achieve a healthy organizational balance.

Developing these muscles has to be intentional exercise. In an experience-driven culture, the notion of testing and experiencing with different approaches, ways of working, and initiatives mix could all contribute to better portfolio choices and balance effectiveness. Critical differentiating qualities are observed from the onset with the initiating processes of portfolio management and how tradeoff decisions are reached early to make the tough choices for what is included in the portfolio mix.

Prioritization skills are quite valuable here too. Most relevant will be how this remains to be a dynamic process across the work of the organization, how it affects budgeting, horizontal working across the organization, and ultimately selecting the right joint strategic metrics to align the behaviors across the organization's culture.

Key Learnings

- Understand the portfolio management ecosystem and how it connects strategy to the ultimate achievement of initiatives' outcomes.
- Get a comprehensive look into the skills that differentiate the effective practice of the portfolio management discipline.
- Learn the key practices needed to align the organization, its leaders, practices, and data.

Creating Experience-Driven Organizational Culture: How to Drive Transformative Change with Project and Portfolio Management, First Edition. Al Zeitoun.

- Learn the art of enhancing effectiveness in reaching the right decisions, fast, while establishing decision ownership.
- Develop an approach to maturing portfolio management that aligns the commitments to that journey.

10.1 The Portfolio Management Ecosystem

For leaders to become effective in portfolio management, they should have an ecosystem view of the portfolio. One way to view the ecosystem is to use Project Management Institute Standard for Portfolio Management, Fourth Edition, and view the portfolio lifecycle as four stages of initiation that need to be done for each portfolio, and then the major stages of planning, execution, and optimization. Of course, with the dynamic nature of the portfolio, the initiatives have to be monitored, controlled, and ideally measured against a set of maturing strategic metrics.

Part of the ecosystem is to think through what is needed to transition to operations, especially from a view of value and who owns the value across the initiative transition. In a discussion with **Ed Hoffman, PMI, Strategic Advisor,** he offered the following views related to the development of the portfolio management muscles:

Creating experiences means having an array of programs' outcome possibilities.

- It's important to basically know the culture's chemistry today and in the future.
- The safest portfolio approach is dependent on a critical mindset and a reflective mindset.
- Portfolio management is about how we want to allocate our time and what we're investing in, within the learning mindset and culture of the organization.
- To make proper portfolio selection is to understand implicitly that a proper portfolio mix is based on the way of laying out possibilities for both the short-term and the long-term for the better of the portfolio outcomes.
- I think selection is an important part of the portfolio process development and maturing within the organization and the learning of its leaders to become strategic thinkers.
- It is part of that notion of understanding the concept of system and then how to make those strategic choices decisions.
- This is also about considering the larger array of possibilities and making decisions based on the "why."
- I think the portfolio view is really the key separating point for organizations' and societies' projects that are successful; it's a connection to ongoing learning in terms of anticipating and predicting capabilities.

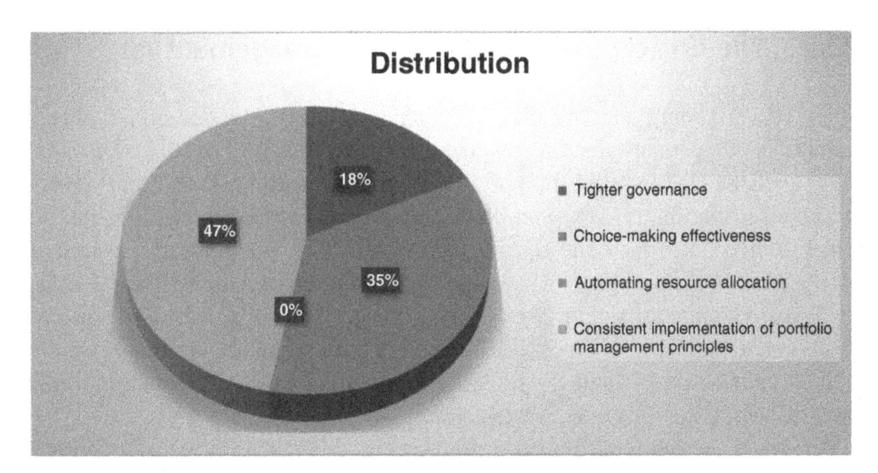

Figure 10.1 Successful Organizational Portfolio. *Note: Based on LinkedIn Open Polling, April 2024.*

- The initiatives' sponsors of the future need to build these capabilities, which will add credibility to the leaders' strategic sense and the proper perception of system understanding.

In testing one more of the eight hypotheses beyond this work, in order to assess the importance of portfolio management across the organizational ecosystem, the following hypothesis was put to the test: "***Choice-making across the portfolio is a strategic responsibility***."

The question used was: "*What contributes the most to the success of strategically managing organizational portfolio of projects?*"

Figure 10.1 shows that consistent implementation of portfolio management principles received the highest score with 47% and is a reflection of the importance of the principles in driving mindset changes and a way to working that has an ecosystem view. If combined with the 2nd highest core, 35%, choice-making effectiveness, which is traditionally a key strategic side of portfolio management principles, the combined score is 82%.

These combined two elements show the holistic nature of portfolio management in driving strategic mindset of the organization and that strategic choices are a muscle that should be developed and invested in by management in a future of work that values curiosity and experiencing.

> **Tip**
>
> Creating an ecosystem view of portfolio management ensures the right strategic emphasis on enhancing the choice-making muscles and portfolio practices' impact.

10.2 The Critical Skills for Portfolio Management

The PMI's Standard for Portfolio Management, Fourth Edition, highlights a number of competency areas, as reflected in Figure 10.2. PMI covers seven capability areas needed to build the portfolio manager's skills toolbox. These seven, as shown in the figure, are: ***portfolio strategic management and alignment***, which is needed to have the ecosystem view of value; ***portfolio management methods and techniques***, which help the leader in structuring the portfolio work and reaching the right priorities and decisions; and the third is ***stakeholder engagement***, which continues to be a differentiating quality, the more responsibility leaders take beyond the project level, as they have to integrate multiple initiatives into the mix of their responsibility.

The 4th is the ***leadership and management skills***. This is naturally critical given the communication qualities these leaders have to possess to work across multiple settings and possibly cultures. The 5th is ***risk management***, which arguably needs to be at the enterprise level to ensure that there is a strategic view taken when the value of the portfolio is being assessed along the lifecycle. The next capability area is ***organizational change management (OCM)***, which arguably creates the necessary edge for portfolio to excel in the cross-organizational change expected as a result of the portfolio work. The 7th is ***systems thinking***. The ability to integrate the various components of the portfolio, programs, projects, and other work is a strategic quality that the portfolio managers and directors have to possess and continue to develop for them to create the proper impact.

Based on the features of the experience-driven culture addressed in this work, the figure also reflects the foundational elements of curiosity, adaptability, resilience, and choice-making, which contribute to the inspiring leading and running of the portfolio teams into the future. These portfolio managers of the future are most suited to operate at the executive levels of organizations. The experiences that they have in the trenches of programs and projects and working across the ecosystem, handling the many tradeoffs and decisions needed, likely qualify these leaders for the top job.

In our PMWJ article, Kerzner and Zeitoun (2023), March, we addressed how organizations benefit from being strategic about how they approach their portfolios of programs and projects and the skills that need to be nurtured for sustaining that discipline and excellence into the next decade.

10.2.1 Introduction

The future belongs to organizations that are able to continually stretch. It has become evident with intense disruptions that the only operating models that could survive and sustain growth in the future are the adaptable and resilient ones.

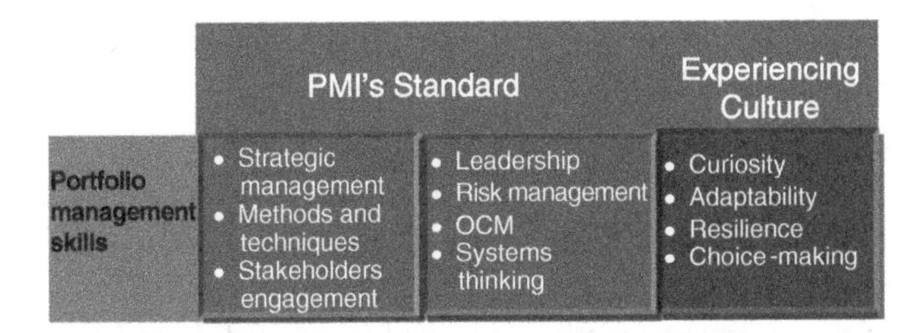

Figure 10.2 Critical Portfolio Management Skills. *Note*: PMI's Standard for Portfolio Management, 2017/Project Management Institute.

Value-focus has been widely discussed in research and practiced for the past few years. In addition, the need to find better and more effective ways to implement strategy has been at the forefront of what executive teams have continued to be concerned with.

We learned a lot over the years about what works and what does not in the utilization of programs and project management. Strategically, organizations have finally started realizing the true value of structured, yet highly adaptable, practices for driving the major change missions they face. The next decade will see a tremendous focus on building these program and project management muscles and how this directly contributes to impacting organizational operations and how they deliver value to their most critical stakeholders.

During the past decade, there has been a great deal published on the business benefits and business value resulting from effective project and program management practices. Companies are now realizing that project and program management are more than just another career path position. They have become strategic competencies necessary for business growth, customer satisfaction, and sustainability of the business.

What most companies fail to realize is the impact at the organizational level and on the company's business model resulting from the elevation of project/program management from a career path to a strategic competency. New organizational structures are being developed and, in many cases, resulting in major changes to the firm's traditional business model.

10.2.2 The Strategic Benefits of Good Project/Program Management

There is an adage that some trainers use in project management courses: "Any executive that always makes the right decisions probably is not making enough decisions." Applying this adage to the project selection process, not all projects approved by senior management will be successful. But what we do recognize as happening in organizations is that the project success rate increases as organizations become good at project management.

When the project success rate increases, there are several changes seen in companies, three of which are as follows:

- Program/project management becomes a strategic competency rather than just another career path position.
- A significant increase in the number and new types of projects the company needs for strategic growth.
- Project and program management practices are applied to all the projects/programs rather than specific categories.

10.2.3 Operating Models of the Future

Forecasting major changes to the operating models of the future could begin with an understanding of the drivers that will necessitate the changes expected to take place.

- Increased need to respond to customers' changing demands and growing expectations.
- Having a delivery model that is reflective of the strategic focus of the business, which has been shifting to delivering change and decreasing investments in traditional operations.
- Need to balance business as well as technical decisions.
- New and more mature metrics to measure what success looks like.
- Use of artificial intelligence (AI) and digitalization to enhance the quality and speed of decisions.

Building new types of operating models that will sustain growth and ensure value delivery will shape the DNA of the future organization. One of the greatest strategic shifts executives have been making is expanding their views of what project and program management muscle development achieves for their organizations strategically. These executives are now holistically seeing how to connect and build the end-to-end value stream of the business on a foundation of proper program and project management practices.

This creates a great future opportunity for the program and project practitioners and the strategic units that own the cascading of the organizational strategy into a prioritized set of initiatives. Operating models in the future are, in essence, seeing a set of key characteristics shine:

- Investing in building an experience-driven culture that uses programs and projects to enrich those experiences.
- Putting digitization in the center as a critical component of delivering initiatives' outcomes.

- Exemplifying new forms of leadership necessary for the next decade of organizational excellence.
- Applying adaptive delivery frameworks, while creating a fine balance between alignment and autonomy.

10.2.4 Organizing of Future Projects

As the number and types of projects increase, companies are being challenged on how best to organize for the growth of projects. For several decades, project management practices were established for managing a single project rather than a group of projects. The single projects were mainly traditional projects that were started with well-defined requirements, a business case, and a statement of work. Other types of projects, such as those related to business strategy, advances in technology, research and development (R&D), innovation, and business opportunities, were managed by functional managers who were allowed to use their own approaches, often not including any of the traditional project management processes, tools, or techniques. The success rate for many of these nontraditional projects was significantly lower than the success rate for traditional projects.

For traditional projects, project managers were assigned from the controlling functional unit. Resources were assigned, when necessary, from other functional units using a strong, weak, or balanced matrix organizational structure. When the project was completed, regardless of whether it was a success or failure, the employee would return to his/her functional unit. The success or failure of the project most often had no impact on the employee's performance review by the functional manager.

The new types of projects, accompanied by the organization's recognition that project management has become a strategic competency, have brought forth new issues, as shown in Figure 10.3.

10.2.5 Strategic Focus Shifts

The strategic nature of many of the new projects is forcing organizations to better understand the alignment between projects/programs and business strategy. Project management is now part of the organization's strategic business plan with a focus on long-term rather than short-term business benefits and business value.

Integrating strategic planning and program/project management will create alignment issues and conflicts resulting from new OCM initiatives. Some of the issues expected to surface include:

The focus will be on managing and organizing groups of projects rather than a heavy focus on individual projects.

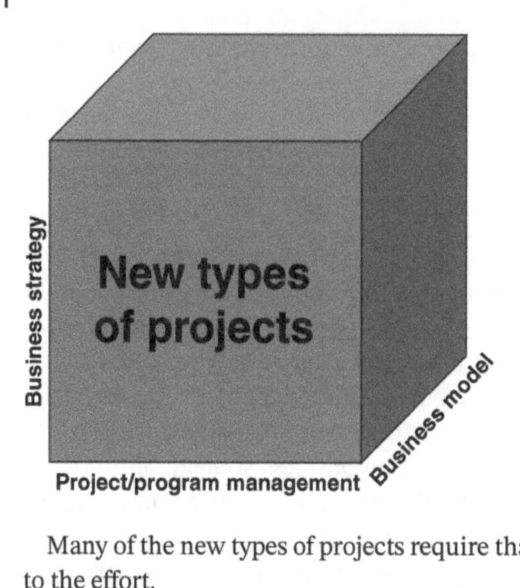

Figure 10.3 Future Projects Alignment Attributes.

Many of the new types of projects require that the workers be assigned full-time to the effort.

- Functional units will be created for managing groups of projects, and employee performance reviews will be the responsibility of the new functional units.
- There will be a much greater level of participation by people outside of the project team and also outside of the organization.
- There will be new problems and issues that will be more complex, unlike what has been seen previously on traditional projects.
- Problem-solving and decision-making will become more complex. Organizations will need additional soft skills training in these areas.
- The sharing and communicating of strategic information will become a necessity for effective problem-solving and decision-making to occur.
- The measurement of success on these new projects will be the business benefits and business value created, rather than just creating a deliverable for a client.
- The focus will be on long-term rather than short-term business benefits.
- Program and project teams may not be self-managed. Organizational units may be created for managing groups of projects, and project management leadership will be horizontal and vertical rather than just horizontal as with traditional projects.
- New business-related tools and techniques will be part of project management practices. Project managers will see themselves as managing part of a business rather than just a project.
- The program manager's focus will be on managing the linkage between the organization's strategic goals and the program's objectives rather than simply creating a deliverable for a client.

10.2.6 The Path Forward

As can be seen from the abovementioned list, program and project management muscles are quickly becoming a strategic/business process with the intent of achieving strategic goals rather than just deliverables on a project or a program. These initiatives' leaders will need to formulate strategies to compensate for changes in the enterprise environmental factors and market conditions.

The good news for tomorrow's operating models is that executives are gradually comprehending the strategic potential of program and project management and thus paying closer attention to the muscle building required. There is now a clear realization that the future operating models need continual transformation and that properly and strategically prepared program and project managers will be essential to driving future successes.

Shifts in how we work and how we deliver value dictate a new mindset and adaptable practices that the program and project leaders are best at bringing to the forefront. Any reinvention of future operating models will be affected by the delivery vehicles of products and services to customers and other stakeholders, and thus the importance of these muscles will continue to grow.

It is with adaptability, creation of experience-driven culture, and the effective use of digitalization that the program and project management practices will continue to demonstrate strategic impact. Empathizing to the growth mindset needs of these future leaders will directly contribute to the effectiveness of the next-generation operating models.

> **Tip**
> The critical skills of portfolio management combine a set of thinking, change, risk, leadership, and experiencing qualities that enable these leaders to sustain value.

10.3 Aligning the Enterprise

Under the experience-driven culture principles, the aligning of the enterprise is linked to the common ethos, joint views of value, and, in the portfolio case, a holistic view of how the success of the portfolio will be measured and connected to strategic value and impact.

Achieving alignment requires a clear cascade of the overall strategic goals using proper strategic approaches and plans. Most of the alignment work is empowered by the safety of experimenting and the allowance for collaboration and communication to create a connected enterprise.

Figure 10.4 The Aligned Enterprise.

As shown in Figure 10.4, the analogy to rail tracks is the targeted goal in achieving cross-team alignment. It is about how the efforts and energy could be streamlined to progress in a clear and consistent direction while maintaining a view of the horizon, as illustrated in the figure. An additional element in this alignment is the speed at which the team operates safely and in harmony while forward movement is achieved across all the organizational moving parts, such as in the case of a train.

In the discussion with **Ed Hoffman, PMI Strategic Advisor**, he highlighted some additional attributes that support achieving this enterprise alignment.

"I believe culture is essential. I think that in the environment we're in right now, culture is the key driver for success within projects or anything else. This has to do with connecting knowledge capabilities that diverse people have, and the culture could either support that to happen or prevent it from happening. So, the right culture encourages the sharing of knowledge, the generation of it, and keeps that promotion of learning and development going. This is the kind of culture that drives getting the right things done across the organization."

Tip

It takes a village to align the enterprise. This is a continual effort to connect the stakeholders to the joint direction of value to be achieved from the portfolio of initiatives.

10.4 Decision-Making Excellence

Peter Drucker once said: "When you see a great business, someone made a courageous decision."

One of the most common important dimensions of portfolio management efforts success is the effectiveness of decision-making. Excelling in decision-making in the future is a finite exercise of balancing people and technology. The advancements in digital fluency addressed in Chapter 7 of this work will directly contribute to the improvements in the speed and quality of decisions. AI-enabled decisions will minimize the likelihood of risks and issues dominating where the project portfolio leaders will be spending their time in the future.

Many of the learnings in the following case study could be ripe examples for the application of advancements in decision-making enablers. In a world that is driven by clock speed, having the safe experiencing and simulating of the real world in the virtual world could become major differentiators for effectiveness of decisions in the portfolios of initiatives, similar to what is highlighted next in the case study.

Case Study

The Berlin Brandenburg Airport[1]

Project managers are trained to expect the unexpected and then to apply risk mitigation strategies when necessary. This type of training is included in most courses and is most often discussed in all areas of knowledge in the PMBOK® Guide, rather than just the risk management section.

On projects that are short in duration or well-defined at the start, the project manager and the team usually have the expertise to mitigate the risks. The project manager may not possess all the technical expertise needed for all facets of the project but usually has a reasonably good understanding of what questions to ask to make a viable decision. If the team cannot make the decisions themselves, then they must rely on the governance personnel and stakeholders for support, assuming governance personnel and stakeholders possess the necessary knowledge. However, even with proper staffing and governance, schedule slippages (hopefully short) and cost overruns (hopefully small) can still occur.

Unfortunately, there are several situations that can lead to large cost overruns and long schedule slippages. A typical list might include:

(Continued)

1 Kerzner, 2022/John Wiley & Sons.

Case Study (Continued)

- Having a technically inexperienced or nonspecialized project team.
- Assigning stakeholders and governance personnel based on politics and union affiliations rather than experience on this type of project.
- Having too many stakeholders, many of them with hidden agendas that conflict with the best interests of the project.
- Approving the project without supporting data to justify the need for the project and the expected benefits.
- Beginning the project with unrealistically low-cost estimates to gain support, such as political support and endorsements.
- Failing to examine best practices and lessons learned from similar projects worldwide.
- Changing governance personnel and stakeholders throughout the life of the project so that senior decision-makers can avoid being held accountable for previous bad decisions.
- Not validating the credentials of team members assigned to critical positions.
- Making unrealistic assumptions about the enterprise's environmental factors and their existence.
- Failing to understand the effect of competitive forces.
- Awarding procurement contracts based on politics rather than capabilities.
- Awarding procurement contracts before the planning documents are finalized.
- Decision-makers have insufficient knowledge about the technology and approve costly scope changes without careful consideration.
- Lack of communication throughout the life of the project between the governance personnel, the stakeholders, and the project team.
- Having a stakeholder and governance team that lacks an understanding of project management and what information should appear in progress reports.
- The lack of a central organization with project management knowledge to oversee the entire project.

Each of these situations can create major headaches during cost and schedule management activities. Now, let us ask a critical question: What would happen if all these situations occurred on the same project?

This is how you convert a US$ 3 billion project into an US$ 8 billion project and incur a schedule slippage of more than eight years. This is the Berlin Brandenburg Airport.

Justifying the Need

Large infrastructure construction projects are not as uncommon as most people might believe. Government agencies worldwide may justify the need for a large, international airport perhaps once a decade, if at all. Any historical data from other similar airports, if it exists, would be outdated due to changes in technology and consumer expectations.

When the Berlin Wall fell in 1989, the newly unified Germany identified a need for a modern international airport that would support the expected growth in air travel. Berlin's three airports were becoming outdated. Tempelhof Airport was the oldest and was scheduled to be closed. East Berlin's Schönefeld Airport was used by low-cost airlines. This left Tegel Airport, a gem of efficiency that opened in 1948, as Berlin's main airport for international travel. The construction of a modern international airport in Berlin would close down Schönefeld and Tegel.

The justification for a new airport required a valid business case. Unfortunately, there were more arguments against building the airport than in favor of it. Several of the major concerns were:

- Will there be enough passenger traffic given that Berlin is not necessarily considered attractive as a "business market"?
- Will any of the airlines other than Air Berlin, which ceased operations on October 27, 2017, due to bankruptcy, consider making the new airport a hub for connecting passengers? The Frankfurt and Munich Airports are already well-established international hubs and would be serious competitors.
- Will the new airport be financed with public or private funds or a compromise of both?
- Will ownership reside with the public or private sectors, or both?
- Given that it may never become an international airport, how do we attract stores for the airport when much of their revenue comes from passengers connecting to other flights? Having too many stores could appear as competition with other international hubs.
- How should we handle noise abatement practices for people living near the new airport and close to runways? Will it be necessary to soundproof some homes? Should the airport prevent nighttime arrivals and departures between 11:00 p.m. and 6:00 a.m.?
- How will we handle relocation efforts for families on land we wish to purchase for the airport?

(Continued)

Case Study (Continued)

Governance Issues

Two issues usually investigated in detail on large infrastructure projects are in-depth discussions of project governance and technical problems. The decision was made to go ahead with the airport fully aware of the risks. However, there were critical discussions as to whether the airport should be privately or publicly owned. The privatization plan was scrapped, and the company that developed the privatization plan was paid more than US$ 55 million for their effort.

The new Berlin Airport would be planned, owned, and operated by Berlin, Brandenburg, and the federal government of Germany as BBF Holding. Shortly afterward, BBF Holding became Flughafen Berlin Brandenburg GmbH (FBB) and remained under the ownership of Berlin, Brandenburg, and the federal government.

With limited project management knowledge residing in the leadership of FBB, an optimistic budget of approximately US$ 3 billion was announced for the airport, along with a launch date of October 30, 2011. As work started, many of the governance situations discussed previously began to appear, in addition to technical issues. The result was now expected to be an airport that would be more expensive and an opening day that would be delayed. BBF would require more funding, necessitating loan guarantees.

In June 2010, FBB announced that the "ambitious" October 2011 deadline for opening the airport could not be met because of, among other things, the bankruptcy of the construction planning company. In January 2013, the opening date was pushed to at least 2014. There were several personnel changes at FBB as well, both voluntary and involuntary. The airport's former technical director was accused of having accepted US$ 680,000 in bribes. In 2014, a public tender was announced for any European company to bid for the planning and construction coordination of the airport. No useful offers were received. In 2015, one of the most important construction companies filed for bankruptcy.

As delays were announced, things went from bad to worse. Airlines and businesses that invested money based on a targeted opening day were suffering from potential cash flow losses and filed lawsuits for financial damages. Even railways filed lawsuits for non-usage of the ghost station below the airport. The man who filed a claim for US$ 60 million in damages suffered by Air Berlin due to the continued postponement of the opening in 2012 and who described the airport as "a huge embarrassment for Berlin, and the whole world is laughing at us now," then went on to become the chief executive officer (CEO) of FBB. Several people were remanded for alleged bribery and corruption activities. In May 2016, a whistleblower on the airport project, who had alerted the public to major corruption within the project, was poisoned with a "deadly substance" but survived after a three-month illness. During the delays, the CEO of the Airport Board was replaced.

Technical Issues

Many of the technical issues were a direct result of issues that the governance team failed to understand. Some of the technical issues were as follows:

- The fire alarm and smoke exhaust system were not built according to the construction permit.
- The fire alarm and smoke exhaust system were improperly planned and constructed, built to exhaust smoke underneath the floor. This violates the laws of physics that say hot air rises.
- The person who designed the fire alarm and smoke exhaust system was an engineering draftsman rather than a qualified engineer. He was later removed from the project, and the entire system had to be rebuilt at a nine-digit-figure cost.
- Some 600 fire protection walls had to be replaced because they were built out of gas concrete blocks that provided insufficient fire protection.
- In response to the fire system mistakes, the airport planned to employ hundreds of nightclub bouncers to sound alarms and open doors to exhaust smoke manually because the automated system was unable to open the doors. Technical issues involving the electric doors became public on January 18, 2017. It was discovered that 80% of the doors would not open, which created concerns around venting of smoke in case of a fire.
- A projected increase in the number of passengers moving through the airport meant that the fire emergency load and smoke control system were inadequate for the main terminal and railway station in the basement and had to be replaced.

(Continued)

Case Study (Continued)

- The underground railway station needed a redesign for the underground part of the fire exhaust system. Incoming or departing trains might suck smoke into the station, so airflow guidance was needed to avoid this effect.
- The 750 display screens that were switched on continuously for more than six years, with no passengers at the airport, were at the end of their useful life and had to be replaced.
- There were 90,000 m of cables incorrectly installed in concrete instead of running through shafts.
- The sand-lime brick used in the foundations of the airport was not sufficiently rated for load, necessitating a costly replacement of much of the underground cabling and reinforced concrete beams.
- Due to water overflow within the cable ducts, 700 km of cable needed to be replaced. The ducts were not leakproof against incoming water and had eroded in the decade since they were first installed.
- Inspectors discovered flaws in the wiring. The cable conduits held too many cables or held cables in incompatible combinations, such as phone lines next to high-voltage wires.
- As late as 2019, deficiencies in the electrical system and wiring continued to be found, with issues in the wiring arrangement and the ability to withstand sustained usage and heat.
- Approximately 37 mi of cooling pipes were allegedly installed with no thermal insulation. This required the demolition of walls.
- Some exterior vents were in improper locations, allowing rainwater to enter them.
- Some lightning rods were missing.
- The backup generator powering the sprinkler system did not provide adequate power.
- Parts of the sprinkler system had sustained failures. The sprinkler heads were replaced for increased water flow, but the pipes were too thin to carry it; as a result, the roof needed to be opened for the pipes to be replaced.
- Motors used to open and close windows would not operate above 30°C, necessitating their replacement.
- There were 4,000 doors that were incorrectly numbered.
- Electric doors had no electricity.
- On March 5, 2017, the transformer station exploded.
- Several escalators were too short.

- There were not enough check-in desks. During real-world testing in the lead-up to the opening, each check-in counter was supposed to handle 60 passengers an hour, but staff members were only able to deal with half as many people.
- Even without cars, the floors in the car park began to sag because they did not contain enough steel girders.
- Thousands of light bulbs were running nonstop since officials could not figure out how to turn them off.
- Hundreds of freshly planted trees had to be chopped down because they were the wrong type.
- The concrete foundation needed to be partly rebuilt.
- Every day, an empty train goes to the unfinished airport to stop the tracks from getting rusty and prevent mold in tunnels.
- The family of Willy Brandt (a former chancellor) requested his name be removed from the airport so as not to be associated with the ongoing embarrassment.
- Flight paths and sound protection zones were incorrectly calculated.
- The emergency line to the fire department was faulty.
- The airport's roof was twice the authorized weight.

Among all the technical problems listed above, perhaps the biggest obstacle to the airport's opening was the serious issue with the automatic sprinkler and fire safety systems. This was a sensitive issue with the German authorities because on April 11, 1996, a fire began inside the passenger terminal of Düsseldorf Airport, killing 17 people. At that time, neither a sprinkler system nor fire doors were mandatory. For this reason, when in the fall of 2011, a team of inspectors (logistics, safety, and aviation experts), known as ORAT (Operational Readiness and Airport Transfer), arrived at the newly constructed Berlin Brandenburg International Willy Brandt Airport; they examined everything from baggage carousels to security gates, giving high priority to the fire protection system. When they simulated a fire, some alarms failed to activate, and others indicated a fire, but in the wrong part of the terminal. The ORAT technicians discovered that high-voltage power lines had been laid hastily alongside data and heating cables, a fire hazard in its own right. In addition, the smoke evacuation canals designed to suck out smoke and replace it with fresh air failed to work properly. The experts concluded that in an actual fire, the main smoke vent might well implode.

Given this critical situation, the CEO and the staff tried to minimize the risks and told the airport's board of oversight and the commissioner, who had the

(Continued)

Case Study (Continued)

final authority to issue the airport the operating license, that they were working through some issues but that everything was under control. The solution was what some called an "idiotic plan." Eight hundred low-paid workers armed with cell phones would take up positions throughout the terminal. If anyone smelled smoke or saw a fire, they would alert the airport fire station and direct passengers toward the exits, neglecting the fact that the region's cell phone networks were notoriously unreliable, or that some of these workers would be stationed near the smoke evacuation channels where fire temperatures could reach 1,000°F (~540°C).

The preparation for an extravagant grand opening continued despite these issues being unresolved. On May 7, 2012, less than four weeks before the scheduled opening, the commissioner refused to grant the airport an operating license. An announcement was made that the airport wouldn't open as scheduled and that two airport company directors, three technical chiefs, the architects, and dozens, if not hundreds, of others were fired or forced to quit or left in disgust. The government was spending US$ 18 million per month just to prevent the huge facility from falling into disrepair.

Even though it may be difficult to understand and discover the real reasons for failure, some evidence is quite clear:

- The contrasting vision between the CEO and the architect. With the construction underway, based on the forecasts for air traffic that would have serviced up to 27 million passengers yearly, the CEO dreamed of making the airport a Dubai-like luxury mall. The rationalization was that airports earn significant income from nonaviation businesses. Therefore, the following proposal cropped up: why not build a second level, jammed with shops, boutiques, and food courts? The architect derided what he referred to as this request for "mallification," but he accommodated the demands. On the same line, there was the proposal to allow the giant Airbus A380 to land at the airport. Even though no airline indicated it wanted to fly an A380 to Berlin, the request was to remove the walls at one end of the terminal so that an extra-wide gate could be built to accommodate it. The architect complained about all these changes and did not hesitate to say that the CEO had no concept, only insatiable demands.
- **The change of the project size**: The terminal dimensions increased from 200,000 to 340,000 m^2 (dwarfing Frankfurt's 240,000 and just shy of Heathrow Terminal 5's 353,000). The work was divided among

seven contractors, and that soon grew to 35 and included hundreds of subcontractors. Several engineering and electronics companies, led by the German giants Siemens and Bosch, struggled to retain control over the complex fire protection system that included 3,000 fire doors, 65,000 sprinklers, thousands of smoke detectors, a labyrinth of smoke evacuation ducts, and the equivalent of 55 mi of cables.

- **Management's stubbornness**: At one point, in 2009, outside controllers advised the management team to shut down construction for half a year to give the architects and contractors time to coordinate efforts. The request was ignored, and just months before the scheduled June 2012 opening, the terminal was a mess.

During the next two and a half years, many of the issues remained unresolved. The new management team started with good resolutions, but soon the contrasting vision between the CEO and the engineering chief appeared again. The feeling was that the CEO wanted to get the airport up and running even with, according to the engineering chief, ill-conceived schemes.

Therefore, in December 2014, the CEO quit, and in February 2015, the board hired the former chief of engineering at Rolls-Royce Germany and a former Siemens manager as his technical director. One of the first moves the two made was to yank out and reinstall the miles of cables. Then they turned to the fire prevention system. Smoke would now channel upward through chimneys, in accordance with the laws of physics.

The news media printed stories about the continuous delays and cost overruns at the airport. Some of the comments were as follows:

- "My prognosis: the thing will be torn down and built anew."
- "The airport, built by a public consortium of two federal states and Germany's federal government, is the world's most expensive white elephant."
- "The expectation that Berlin would be able to establish itself as an international and even intercontinental aviation hub was always utterly unrealistic."
- "As there is also a very small local business travel market, especially for long-haul, no airline would be able to turn Berlin as a mega hub into a viable business case. There is simply not enough premium-class traffic demand to and from Berlin."
- "The clients were tripping over each other with requests for changes."
- "The people responsible for technical oversight were saying, 'We cannot do this within this amount of time,' but the CEO would answer, 'I don't care.'"

(Continued)

Case Study (Continued)

- "The number of defects that they've found has grown to 150,000."
- "It has been a long, difficult road until the final approval from the building authorities."
- "You have to say that it is a really cool airport. The architecture is good. The concept is good. It is very easygoing and easy to navigate. It should please a lot of people if it ever gets finished."

On October 31, 2020, Berlin's newest airport finally opened to the public, nine years after its original launch was planned. Two A320neo aircraft landed to mark the milestone opening; one was operated by easyJet, the other by Lufthansa. When the two planes landed, they were welcomed with a huge fountain display.

Value-in-Use

Large infrastructure projects are notorious for cost overruns and schedule slippages. The larger the project, the greater the variance from the original estimates. However, this does not mean that the project cannot be regarded as a success. Denver International Airport had similar issues. The original budget was US$ 1.2 billion, and the final cost was $5 billion, and late. The Opera House in Sydney, Australia, was 10 years late in completion and 14 times over the original budget. Both projects are now seen as successes.

Regardless of the magnitude of the cost overrun and schedule slippage, the real definition of success for these types of projects should be measured by value-in-use after launch. With expansion plans being developed, the airport is expected to handle 55 million passengers by 2040.

Despite being infamous for thousands of defects during construction, the airport will open and most likely be a success. For the Berlin Airport, it may take several years to identify the real value-in-use from the project. In other words, we could say: result are what matters. However, the lessons learned and the analyses behind such case studies will help, hopefully, in the future to avoid mistakes, so that not only will the result be something good but also something obtained with efficiency from a management point of view, and above all without squandering public funds.

Management Analysis and Recommendations

In a case like the one of Berlin Brandenburg Airport, it is really difficult and risky to give some opinions about what could have been done better, who are the responsible managers, and what did not work from a management point of view. Even for a court, which can have access to all documentation and can interrogate the involved people in a project, it is hard to identify and sentence a responsible person. In the case of the CEO fired in 2012, for instance, he sued for wrongful termination, and in late 2014, a Berlin court ordered the airport owners to pay US$ 1.28 million in damages for his dismissal, saying the board of oversight shared responsibility for the fiasco!

Many observers underline that for the Berlin Brandenburg Airport, there was a clear inaccuracy in budget estimation, causing management problems all the way through execution. In addition, they recognized politics as responsible for this mistake, which plays a significant and influential role in public works. In general, the wrong attitude with political leadership is to keep the estimated costs of the construction of a public work low to have more chances to obtain support for the projects, deliberately veiling the potential risks. Therefore, cost overruns rarely come as a surprise.

What seems to be the clear source of problems in the case of Berlin Brandenburg Airport is management's stubbornness, intended as the will to stay on the schedule regardless of the reasonable choices, and the continuous variations to the project. It is important to demand the best from the team to respect the deadlines, but not by sacrificing quality. As for the variations in the project changes, considering that the airport is not something easily changeable, the requests for modifications must be carefully evaluated with an approach to ensure cost-effectiveness.

Based on the abovementioned considerations, it is quite apparent how complex and insidious is the environment in which a project manager operates when it comes to public works: political pressures, requests to stay on the schedules under any circumstance, and scope changes as well. What advice can be given in this extreme work environment? It is undeniable that we must be able to convince people to accept some unpleasant choices, like delays, or to discourage people by considering project variations with big impact and risk.

However, bearing in mind that numbers speak louder than words, it is advisable to collect all the needed information in order to perform accurate risk management analyses to explain to all the stakeholders and (especially) to the superiors the risks connected with some questionable choices. We could say that in such a complicated environment, risk analysis is a powerful tool in the hands of the project manager to be used to persuade in order to avoid inappropriate choices.

(Continued)

Case Study (Continued)

Questions

1 What were the concerns before the construction of the Berlin Airport?

2 What was the main technical issue responsible for the delay, and how did the managers try to solve it?

3 What was the main contrasting view between the CEO and the architect? And who's viewpoint finally won out?

4 Why did managers refuse to shut down the airport to improve the coordination efforts?

5 According to the experts, what were the main reasons for delays and cost overruns?

6 What is most likely the ultimate parameter to judge if a project is a failure?

7 What recommendations can be given to the project managers based on the experience gained from this case study?

References

Wikipedia. (October 21, 2024). Berlin Brandenburg Airport. https://en.wikipedia.org/wiki/Berlin_Brandenburg_Airport.

Delahaye, J. (2020). Berlin's new airport has finally opened after nearly a decade of delays. Mirror, November 4.

Wikipedia (October 22, 2024). Düsseldorf Airport fire. https://en.wikipedia.org/wiki/D%C3%BCsseldorf_Airport_fire.

Hammer, J. (2015). How Berlin's Futuristic Airport Became a $6 Billion Embarrassment. Bloomberg Businessweek, July 23.

James, B. (2018). The Sad Tale of Berlin Brandenburg Airport. One Mile at a Time, April 27.

Kerzner, H. and Zeitoun, A. (2023). Building program and project management muscles: the key to excellent operating models. *PM World Journal* XII (III).

Project Management Gone Bad: The Berlin Airport Project. Project Management in Action blog, November 4, 2018.

Ros, M. (2017). The ongoing saga of Berlin's unfinished airport. CNN, December 5.
Scally, D. (2018). Top Lufthansa boss says new Berlin airport will probably never open. Irish Times, Mar 20.
Schuetze, C.F. (2020). Berlin's Newest Airport Prepares for Grand Opening Again. New York Times, April 29.

Tip
In leading the work of such a complex portfolio, as the case of the Berlin Airport, there are direct benefits to an enterprise alignment that is anchored in experiencing.

10.5 Portfolio Management Maturity

Portfolio success is higher with a clear connection to a purpose that matters. Many of today's challenges in achieving portfolio management maturity stem from gaps in strategic clarity. There are many methods to consider when charting a roadmap for maturing portfolio management practices. Most of these approaches tend to have levels similar to the ones highlighted in Figure 10.5.

Level 1 is typically about showing some initial evidence that the organization and the portfolio teams understand the necessary processes and have begun to use them. **Level 2** is already a great step in the right direction where portfolio standards are widely and consistently followed across many of the portfolio components' teams. **Level 3** is already an advanced stage of maturity, where the enterprise's approach starts being controlled and shows consistent management of the portfolio components. **Level 4** is a confirmation of the progress in level 3 into the strategic alignment of the organization around the ways the portfolio management processes are used to run and change the business. Ultimately, level 5 is about

Figure 10.5 Possible Maturity Levels.

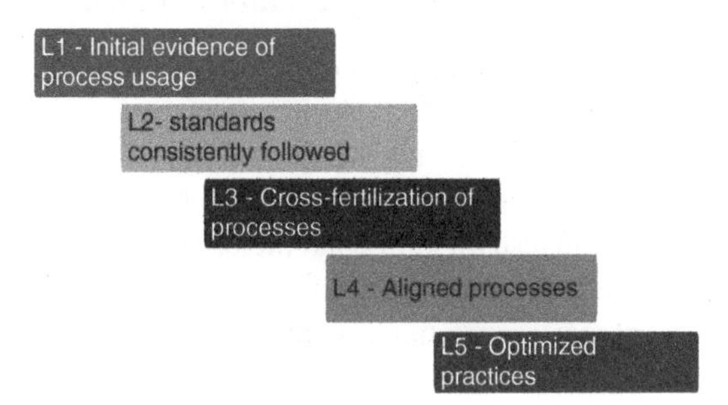

the required continuous improvement that leads to sustaining these experiencing practices that will continue to dominate future cultures.

The key next measuring cascade from these levels would be how to assess the movement in each of these levels toward what is expected in reaching the given level. It is also useful to have a clear description of how to choose necessary improvements to progress across each of these levels. This description of the improvement work could be customizable depending on the industry, the complexity of the portfolio, and a few other technical, and cultural dimensions.

In discussing multiple cultural and portfolio best practices with **Paul Jones, Fujitsu, Head of PMO**, he shared multiple insights that could directly contribute to enhancing the movement on the path of portfolio management maturity.

10.5.1 The Maturing Culture

- Organizations that seem to have a culture that everybody gets on board with recruit people who will fit the culture, or people understand that they will need to change when they join the company. I've heard Amazon is looking at people who they can assess and confirm have the right mindset and the right cultural values to fit in the company. Whether this is good or bad, no one can confirm, yet it won't necessarily be perfect without more work.
- Culture is about people.
- People know what happens in the organizations, and what the company's leadership thinks.
- Culture then becomes a reflection of how leaders behave and, ultimately, the way we treat each other.
- A supporting culture could be shooting to create an A-Team with leader doing something different to be successful in building that mix.
- In the opposite case, and in treating people in a way that they shouldn't be, this starts eroding the organizational culture because people know they need a different culture to successfully contribute to business.

10.5.2 The Supporting Cultural Attributes

- I'm sure consulting organizations will be able to sit down and say, Oh, this is what a good culture looks like.
- I think it's completely unique to a given organization and team.
- It is a unique setting for the right people who want to come to you, to work for you, and also for your customers. This is why they want to partner with you.
- The idea that you can create a boilerplate for some kind of culture that you can apply to organizations is not correct.

- Culture has to fit the organization and its context right.
- There are potentially universal attributes that would create a strong culture, like trust that has to be there.
- What you end up doing is writing down a description of a good person.
- We need to be empathetic; we need to be trustworthy; we need to be respectful.
- This is like the things you'd like to see in a friend, in someone who you actually want to hang out with, someone who you respect.
- These easy emotional type things could build a culture where we are honest.
- Every organization should want to be open, honest, respectful, and empathetic.

10.5.3 Portfolio Muscles Building

I'm seeing that as a change portfolio in your business. Whatever is in your product development portfolio, it could be your transformational portfolio, the things you want to do, and the things you don't want to do. Success in the projects could actually be in canceling the project and that should be seen as a success.

- This requires faith in order to stop projects, and most project managers don't have that because most organizations will still see that as a failure to stop a project.
- As a project manager, to actually say we need to stop this project, if needed, means you are doing your job.
- For a given portfolio to have the right stuff right, your culture has to define success.
- The cultural behaviors cascade down to the portfolio level.
- Just like the example of the Bill and Melinda Gates Foundation and the failures that they had first, they openly talked about how they stood up and celebrated failure.
- The reality is that they weren't celebrating failures, but they were celebrating how they responded to failure.
- A culture of learning is about having the courage to say we need to stop doing this.
- There is a leadership side to this, as many organizations use words and do not translate them into actions.
- It's easy to say we focus on our organization's culture, yet how many times have we seen that in the news? It turns out that they are not necessarily great cultures.
- Culture should be about rewarding the right behaviors; it's how organizations treat those who do not show the right behaviors.
- As an example, when we look at the sales teams in big organizations, where you know the sales guys might ignore others' views or follow the process, then at the end of it, they get rewarded for signing the big deal.

- In examples where there is no proper transparency, leaders could hide information or not address emerging customer issues, resulting in eventually having to get rid of who has covered up the delays or other critical issues.
- Culture is the way we behave under pressure.
- Not having the proper cultural support for portfolio work would cost the company a lot of money and could cost a lot of jobs.
- Companies should use coaching to set up the right culture examples for how to support portfolio management decisions.

Tip

In committing to achieving maturity, organizations need to strategically align their processes, leadership, and governance practices toward achieving portfolios' value.

Review Questions

Parentheses () are used for Multiple Choice when one answer is correct. Brackets [] are used for Multiple Answers when many answers are correct.

1 What is central to the success of developing enterprise portfolio management muscles?
( ) Making sure that this relies on AI copilots.
( ) One standardized approach that fits all.
( ) Being a dynamic process that works across the organization.
( ) No concern about maturity.

2 What are some of the potential contributors to strategic management of portfolio success? Choose all that apply.
[ ] Initiating portfolio work fast.
[ ] Should not link to budgeting cycle.
[ ] Choice-making effectiveness.
[ ] Consistent implementation of principles.

3 What is a possible portfolio competency that directly supports the handling of politics and unaligned views?
( ) Technical expertise.
( ) Less interest in being curious.
( ) Stakeholders' engagement.
( ) Ensure full automation of decision-making.

4 What is an example of major changes in future operating models?
( ) Use of classical metrics in measuring portfolio success.
( ) Slowing need to reacting to customers' demands.
( ) Use of AI and digitalization to enhance the quality and speed of decisions.
( ) Ensuring the growth of controlling leadership.

5 What could be key skills for the success of the portfolio manager's role? Choose all that apply.
[ ] Adaptability.
[ ] Multiple languages.
[ ] Systems thinking.
[ ] OCM.

6 What is a key differentiator for decision excellence?
( ) More time is spent on decision analysis.
( ) Being able to use data to expedite decisions and enhance their quality.
( ) One decision-maker.
( ) Following an exact roadmap for achieving decisions.

7 What is a portfolio management maturity level that shows consistent linkages to strategy?
( ) Standards are consistently followed.
( ) Aligned processes.
( ) Optimized practices.
( ) Portfolio lifecycle.

11

The Adaptable Future Organization

The future organization is adaptable. The ways of working now and into the future will require adaptable ways of working, mindset, and leadership style. With the continually increasing pace of change, the demands on efficiency, and the accelerated delivery speed, this muscle will continue to be a priority.

Top management of organizations has got to take ownership of creating such future cultures. Leading organizations the same way as five years ago, or a year ago, without adjusting to the realities of tomorrow, will likely have higher consequences than at any time in previous history. With the data revolution the world lives in, there is no hidden information from customers, no barrier to co-creation, and no gap in time between an event and its consequences. The merging of the real and virtual world will only increase the pace of creativity, development, and the expectations around quality, safety, and sustainability targets.

Key Learnings

- Understand how to sustain progress within the changing dynamics in future organizations.
- Learn how to embed the adapting value into the DNA of the organization.
- Explore the principles of achieving the balance necessary to remain responsive to the need for change, while achieving proper level of stability for the projects' portfolio teams.
- Understand the approach and value of co-creating execution plans in the future.
- Develop the muscles and mindset shifts necessary for building organizational strategic planning and executing consistency.

11.1 The Dynamic State of Tomorrow's Organizations

It is expected in this experiencing nature of tomorrow's work that future organizations will remain in a dynamic state. As much as this could mean change

Creating Experience-Driven Organizational Culture: How to Drive Transformative Change with Project and Portfolio Management, First Edition. Al Zeitoun.

fatigue and that humanely this could be burdensome, it creates a differentiating opportunity for humans in the highly digital age. These future leaders' roles will be change-centric and they will be recognized for their ability to lead with resiliency in that future state of the organization.

In an interview with **Ayman Badr, World Health Organization (WHO), Head of the Project Management Office (PMO)**, Ayman highlighted many of the qualities of this future culture and how he sees from his experiences, a few of these future organizations' attributes shaping up to be.

"I'm going to base my views on my experience working across the UN international organizations and it's framed in that perspective, but I think it can be applicable to other organizations. I think when there are organizations that are very rigid and have very firm institutional structures, organigrams, and so forth, culture ends up being the main differentiator. This helps to foster things like collaboration, alignment with strategies, and then also break down barriers that exist, which then becomes a key element in delivering institutional success.

One thing that we do well for the culture to be successful, is design the organization focusing on the few things that we can replicate across the organization, to be a ripple in the notion of culture, or could be a splash. So be intentional at the very beginning of shaping the excellence culture, because as new individuals join the organization, we either want them to conform to the culture we're trying to protect, or we want them to come in and introduce elements that we're looking to change and so being intentional about what/who we bring in, and why, is a big factor."

"In the WHO, where I work, at the end of the day we all serve someone in a community, or serve someone at a headquarters level, yet ultimately those roll down to serving public health at a national level, or a country, so if we can engineer a way where every single individual within the organization understands their service contribution, and how it aligns and links and contributes to the final end goals, then we've achieved a big cultural enhancement.

Learning can happen from the people who work in pockets of the house, who might understand how their efforts end up making impact. Having that kind of connection is strong in understanding what it is to be of service. This helps in reminding us that we are here to be of service to somebody, and not have that pure kind of ego type mentality, but rather a service leadership, the servant leader type mentality, then also have that permeated throughout perfecting how we work."

11.2 The Future Organizational Attributes

In our work in the Project Management World Journal, we tackled the topic of project management best practices that are going to be critical for that dynamic future state. This was published by Kerzner, H. and Zeitoun, A. (2021). Capturing Project Management Best Practices; PM World Journal, Vol. X, Issue XII, December.

Abstract

The strategic goal of all companies is sustainable and profitable growth which includes improving the firm's competitive position, achieving efficient utilization of resources, and maximizing the return on investments (ROIs) in product development. Unfortunately, many companies take a rather passive view on how strategic goals can be achieved and believe that doing business the same old way will continue to be successful in the future. As stated by Albert Einstein,

> Insanity: doing the same thing over and over again and expecting different results.

Companies must understand how project management best practices can elevate their levels of business success and recognize the need for continuous improvements supported by an investment in the discovery of best practices.

Most organizations today realize they are managing their business by projects. Project management is no longer restricted to a few functional domains but is now utilized throughout the organization. Today, project management is viewed as a strategic competency needed for corporate growth rather than a career path. Project management is becoming the vehicle needed to capture, evaluate, and implement the best practices needed for continuous improvements in the firm's performance.

Unfortunately, many companies have a relatively poor understanding of the steps needed to effectively capture and employ best practices. The intent of this paper is to provide a roadmap for how project management can maximize continuous improvement efforts utilizing best practices and strengthen the value and relevance that project management brings to the organization's future.

11.2.1 The Importance of Project Management Best Practices

Project management is no longer regarded as merely a set of activities consisting of a set of processes, tools, and techniques necessary to create deliverables. Project management is now regarded as both a project management process and a business process. Therefore, project managers are expected to make business decisions as well as project decisions. Most enterprise project management methodologies today are integrated with business processes, whereas historically they were just used on traditional projects that were reasonably well-defined. The necessity for achieving project management excellence is now readily apparent to almost all businesses.

Achieving project management excellence is a major component of strategic planning activities. Project management and business thinking are no longer separate activities. Project management is not just about processes; it is now aligned with the delivery of business benefits and value as indicated in the 7th edition

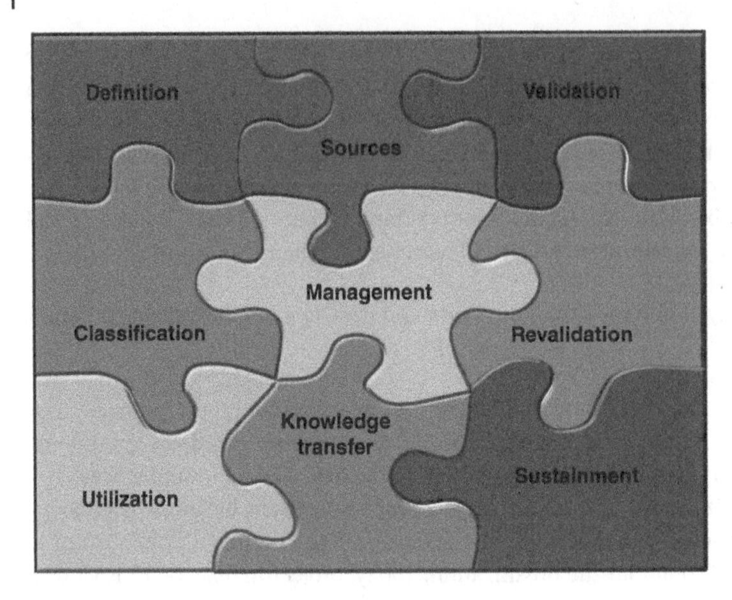

Figure 11.1 Project Management Best Practices Activities.

of A Guide to the Project Management Body of Knowledge (PMBOK® Guide), the flagship publication of the Project Management Institute and a fundamental resource for effective project management for all industries.

One of the benefits of performing strategic planning for project management is that it identifies the need for capturing and retaining best practices, not only for project management applications but for business usage. Unfortunately, this is easier said than done. One of the impediments, as will be seen later in this paper, is that companies today are not in agreement on the definition of best practices and how to manage them. Moreover, companies may not fully understand that best practices can lead to continuous improvement, which further generates additional best practices. Many companies also do not recognize the value and benefits that can come from best practices.

The importance of continuous improvement efforts makes it clear that the discovery and use of best practices should not be left to chance. Historically, most companies never investigate all the necessary steps to determine if a practice is the best approach. Companies are now developing roadmaps for best practices as seen in Figure 11.1.

The activities in Project Management Best Practice activities (Figure 11.1) answer the following nine questions:

1. What is the definition of a best practice?
2. What are the sources where we can discover best practices?

3. Who is responsible for validating that something is a best practice?
4. What are some of the ways to describe levels or categories of best practices?
5. Who is responsible for the administration of the best practice once approved?
6. How often do we reevaluate that something is still a best practice?
7. How do companies use best practices once they are validated?
8. How do large companies make sure that everyone knows about the existence of the best practices?
9. How do we make sure that the employees are using the best practices and using them properly?

In addition to improving performance, capturing best practices can identify ways to eliminate waste, improve the accuracy of estimating work, win new business, enhance the firm's reputation, and help sustain the survival of the firm.

11.2.2 Definition of a Best Practice

A best practice is most frequently defined as a method, technique, or process that is considered superior to other ways of performing the same things and provides the desired outcome with fewer problems and unforeseen complications. The best practice then becomes the standard way of performing certain activities.

Although this definition is commonly used, it is often poorly understood because of the vagueness from use of the subjective term "best." Project management best practices appear as forms, guidelines, templates, and checklists that identify a more prudent course of action. The word "best" implies that there may not exist any other way to perform it better. This is faulty reasoning.

Instead of calling it "the best," a better expression might be a good practice, promising practice, better practice, effective practice, smart practice, or proven practice. Regardless of what we call it, we should follow the steps in Figure 11.1, to show that it has been proven to be more effective than current practices before integrating it into processes so that it becomes a standard way of doing business.

Another argument is that the identification of a best practice may lead some to believe that we were performing some activities incorrectly in the past, and that may not have been the case. This may simply be a more efficient and effective way of achieving a deliverable. Another issue is that some people believe that best practices imply that there is one and only one way of accomplishing a task. This also may be a faulty interpretation.

Definitions of a best practice can be highly complex or relatively simplistic. Companies must decide on the amount of depth to go into the best practice. Should it be generic and at a high level or detailed and at a low level? High-level best practices may not achieve the efficiencies desired, whereas detailed best practices may have limited applicability.

11.2.3 Sources of Best Practices

Best practices can be captured either from within your organization or external to your organization. Internal best practices are usually the easiest to identify but somewhat limited. Internal best practices can be related to the metrics you use, ways to eliminate obstacles to project success, the type of flexible methodologies, the selection of projects, performance reporting practices, working with virtual teams, and the relationships with certain stakeholders. Not all best practices apply to every project. Some best practices are unique to the situation such as in projects related to innovation where the requirements may be loosely defined.

Some organizations rely on the project managers to capture internal best practices and document the findings. This approach does not often work well because of several factors. First, project managers may not be trained in techniques needed to extract best practices. Some companies have professionally trained facilitators that serve this function.

Second, more best practices may be found from project failures rather than successes, but team members are reluctant to discuss what they learned from a failure due to fear that it may reflect poorly on their performance reviews, even though it is in the best interest of the company. Professional facilitators are better trained in extracting information under these circumstances.

Third, project managers prefer to extract best practices at the end of the project and may miss opportunities to gather feedback and insight from team member(s) who completed their assignment before the project came to an end. Seeking out best practices at frequent intervals, such as at gate review meetings, may be best.

There are situations where companies identify the need for project management best practices in a certain process and encourage people to identify possible continuous improvement techniques. As an example, this often occurs when companies seek out better ways to select the best projects for the portfolio while overcoming the barriers due to internal politics, fear of a cultural change, and a poor project prioritization system. There are also barriers related to the unwillingness of people to cancel certain projects that are squandering resources and providing no value.

Another situation is when companies realize that traditional financial measurements of project performance, including use of the earned value measurement system (EVMS), are insufficient in determining the firm's long-term success possibilities. Other project portfolio measurement metrics will be needed.

External best practices come from publications and benchmarking. Publications such as PMI's PMBOK® Guide are seen as generally accepted best practices but may not contain sufficient detail for effective use without some degree of customization. Project management documents, such as the PMBOK® Guide, are limited to project management processes, tools, and techniques, and rarely discuss best practices related to the business side of projects. Emphasis in these documents discusses the forms, guidelines, templates, and checklists that can affect the execution of a project. There are also published articles and graduate-level theses that contain project management best practices information.

Benchmarking is another way to capture external best practices, possibly by using PMO as the lead for benchmarking activities. The two most common forms of project management benchmarking are operational benchmarking and strategic benchmarking. Operational benchmarking focuses on the processes, tools, and techniques used by companies for project execution. Strategic benchmarking focuses on the business side of project management including areas such as project selection, customer satisfaction, stakeholder relations management, and activities related to the strategic planning for project and program management.

Information can also be obtained at seminars, symposia, and conferences that discuss project management best practices. However, experience has shown that most people attend these sessions with the expectation of obtaining information that they can specifically relate to their firm and most presenters provide only generic information.

Companies are often fearful of providing critical best practices information in books or symposia if they believe that the use of these best practices gives their firm a competitive disadvantage.

A case in point: an automotive subcontractor was invited to contribute information for a book on project management best practices. The employee who was invited to provide the information had planned to describe the processes the firm used to minimize the risks when handing off engineering documents to manufacturing team for the start of production planning. Several forms and templates were used to minimize the risks. The corporate legal group declined to give permission to release this information for the book or for seminars and conferences, stating the company spent $3 million developing and testing these forms and templates and saw no reason they should be given to everyone who purchases an $85 book.

Perhaps a more productive form of project management benchmarking is face-to-face discussions between two companies. Both companies must understand that this is a give-and-take discussion and be willing to provide some of their best practices.

The challenge with face-to-face discussions is deciding which company to approach. An executive in the aerospace and defense division of a Fortune 500 company decided to benchmark **against their competitors in the aerospace and defense industry**. After the industry benchmarking study was completed, all of the employees in the company began congratulating one another on the apparent success of their project management activities.

The company then benchmarked itself **against companies considered best-in-class in project management, but none were in the aerospace and defense industry**. This study showed that the company was relatively poor at

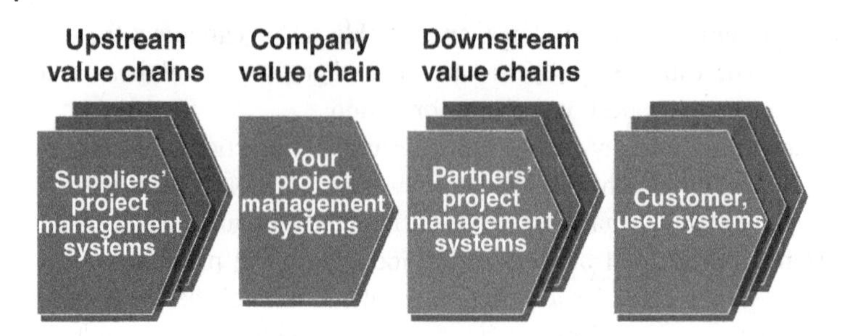

Figure 11.2 The Generic Value-Added Chain.

project management. Senior management recognized the need for more continuous improvement efforts. The leader of the benchmarking study was promoted to Vice President of Innovation and his job was to benchmark and implement best-in-class global project management practices for his firm. Deciding which companies to benchmark against is difficult.

An excellent source of benchmarking information can be found with companies that are part of an organization's value-added chain, as shown in Figure 11.2.

Companies in the value-added chain are upstream and downstream contractors that may become strategic partners and are willing to share best practices information in hopes of a long-term relationship or partnership. Alignment of your project management practices to those of your contractors and distributors can yield fruitful results.

Most of the sources of project management best practices are researched in the forms, guidelines, templates, and checklists used in project management. But as projects become larger and last longer, there is emerging need for behavioral best practices. As stated by Alam et al. (2010):

> A survey conducted in 2010 indicated that about 10% of project problems were related to technical processes and the remaining 90% were related to soft skills management.

The human side of project management may very well yield more best practices in the future than the technical side.

The rapid growth in identifying best practices related to strategic decisions has brought with it an inherent fear among some workers that they may be removed from their comfort zone. Some best practices may require changes to a firm's business model that may result in closing unprofitable product lines/activities or forcing workers to learn new ways of performing work. Project management maturity models of the future are expected to include assessment instruments on best practices processes.

11.2.4 Validation of Best Practices

There must exist a process by which an idea or discovery can be labeled as a project management best practice. Years ago, project managers were allowed to make these decisions by themselves even though they had limited knowledge about best practices. The best practices they approved were constrained by the type of project they worked on, and were limited to improvements to the forms, guidelines, templates, and checklists used for project execution.

Executives began to recognize that they were managing their business by projects. Project management and business-related decisions were interconnected. The best practices discovered and approved by the project managers were project execution oriented and had very limited use for strategic decision-makers. Best practices were lacking for determining the strategic direction of the firm and evaluating and selecting individual projects for consideration as part of the portfolio of projects.

Executives traditionally maintained a hands-off approach to project management activities and relied upon information on time, cost, and scope in performance reports for the decisions they had to make. With the introduction of the 4th edition of the PMBOK® Guide in 2009, projects were focusing on multiple, competing constraints rather than the traditional triple constraints of time, cost, and scope. Also, the enterprise environmental factors, which generally had a limited impact on traditional projects that were well-defined at project initiation, were changing.

The business environment is no longer stable and is now defined by Volatility, Uncertainty, Complexity, and Ambiguity (VUCA). The meaning of VUCA components can change from industry to industry, company to company, and possibly project to project. In a project management environment, VUCA can be described as:

- **Volatility**: An understanding of changes that can occur, usually unfavorable changes, and the forces or events that might cause the changes. The changes could be the need for additional time or funding, poor quality, or inability to meet specifications.
- **Uncertainty**: An understanding of the issues and events that might occur but being unable to accurately predict if they will happen, and when. Uncertainty could be knowing that you need more money or more time, but not knowing how much.
- **Complexity**: An understanding of the interconnectivity of the events discussed under volatility and uncertainty, and any relationships between them. The greater the number of possible events, the greater the complexity. The team can become overwhelmed with information such that they are unable to decide

upon a course of action. Complexity might be knowing that time is money, but not knowing the exact relationship between them.

- **Ambiguity**: Not being able to fully describe the events or misreading the risk events that can affect the outcome of the project. This may be the result of a lack of precedence, the existence of haziness, or mixed meanings concerning the events. This occurs when we are dealing with "unknown unknowns." As an example, ambiguity occurs when we do not fully understand the individual events that may cause budget and schedule issues.

The emergence of the VUCA environment, the introduction and use of competing constraints, and the need to integrate project management and business decisions increased the importance of capturing business-related best practices. Executives now believe that PMO, rather than project managers, should take the lead in recognizing, implementing, and managing the expected growth in project management best practices. Articles such as those by Andersen et al. (2007) and Chen (2015) identified the benefits of using a PMO for project management benchmarking. The PMO also had to make sure that decision-makers throughout the company understood the best practices.

There are several types of PMOs that could take on this responsibility. One of the reasons PMOs have this responsibility is because impact studies like those of the Gartner Group (2000) show that companies with a well-functioning PMO will experience half the time and cost overruns as those without a PMO.

As executives were deciding where to assign the best practices responsibility (i.e., which type of PMO), other project management issues began to surface, including:

- Working on too many projects that provided little or no business value, thus wasting resources
- Capacity planning issues that created an imbalance in the assignment of resources to projects and forced project managers to compete for qualified resources
- Lack of metrics and techniques to predict and evaluate project success
- No application of lean project management practices resulting in excessive waste
- Projects not aligned to strategic business objectives
- Project decisions being made without consideration of the company's long-term strategy

Companies began assigning the best practices responsibility to the Portfolio PMO. For many companies, the use of a Portfolio PMO is a new concept. The benefits of a Portfolio PMO include:

- Improved governance on projects
- Better project prioritization efforts

- Faster and better decisions especially related to project selection practices
- Accelerated improvements in project management processes
- Better resource utilization
- More effective risk management and risk mitigation
- Projects aligned to strategic business objectives
- Quicker cancellation of failing projects
- Decrease in project failure rates and an increase in time-to-market

Even though the primary concern of the Portfolio PMO was the control of the strategic projects in the portfolio, the PMO focused on capturing all best practices that could lead to improvements in the firm's long-term success strategy. Some PMOs developed templates and criteria for determining whether an activity qualifies as a best practice if it has a positive impact on the firm's business. Criteria that can be included in best practices validation templates are:

- The best practice is transferable to many projects
- Enables efficient and effective performance that can be measured (i.e., can serve as a metric)
- Enables measurement of possible profitability using the best practice
- Allows an activity to be completed in less time and at a lower cost
- Adds value to both the company and the customer/client
- Can differentiate us from everyone else
- Helps to avoid failure and if a crisis exists, helps us to get out of a critical situation

The Portfolio PMO becomes the custodian of the firm's best practices.

11.2.5 Classification of Best Practices

As the number of best practices increases, companies have found the need to create a best practices library. The categories within the library can range from general to specific applications and, as expected, the quantity of best practices can vary. Each company may have its own unique classification system.

An example of best practice classification levels might be (from general to specific):

- Professional standards
- Industry specific
- Company specific
- Project specific
- Individual usage

Best practices can be extremely useful during strategic planning activities and levels may be identified to support various strategic planning activities. The bottom two levels described earlier may be more useful for project management strategy formulation, whereas the top three levels are more appropriate for the execution or implementation of a strategy.

11.2.6 Management of Best Practices

There are three players involved in the management of the best practices:

- The Portfolio PMO
- The best practices library administrator, who may reside in the PMO
- The best practice's owner

The PMO usually has the final authority in the identification of a best practice and the placement in the best practices library. The PMO also decides how frequently each best practice may need to be updated, whether it should be removed from service, and whether restrictions should be placed on who is allowed to view certain company-sensitive best practices such as those related to strategic decision-making. Some best practices libraries require viewers to use their employee identification codes for access to certain best practices. The PMO may take the lead in project management benchmarking activities and the selection of which companies to benchmark against.

One person assigned to the PMO usually functions as the library administrator. One component of the administrator's responsibility is to track how frequently certain best practices are viewed. This may indicate the interest in this best practice and how frequently it should be reevaluated.

The administrator is also responsible for ensuring that the library is correctly positioned in the firm's knowledge repository. As seen in Figure 11.3, integrating project management practices with ongoing business needs and strategic planning activities can bring forth a significant number of best practices. As such, best practice libraries become part of the firm's knowledge repository or information warehouse used for decision-making.

As the number of best practices increases, the PMO may not be able to manage all the best practices. Companies have created the position of best practices owners to assist the PMO with continuous improvement and usage of the best practices. The owner of the best practice usually resides in a functional area and may have a dotted reporting line to the PMO which has the responsibility of maintaining the integrity of usually one best practice. Being a best practice owner is usually an uncompensated, unofficial title but is a symbol of prestige. Therefore, the owner of the best practice tries to maintain continuous improvement of the best practice.

11.2.7 Revalidation of Best Practices

Best practices do not remain best practices forever. Continuous improvement efforts require that best practices be reevaluated periodically for applicability because best practices may lose their value if allowed to age.

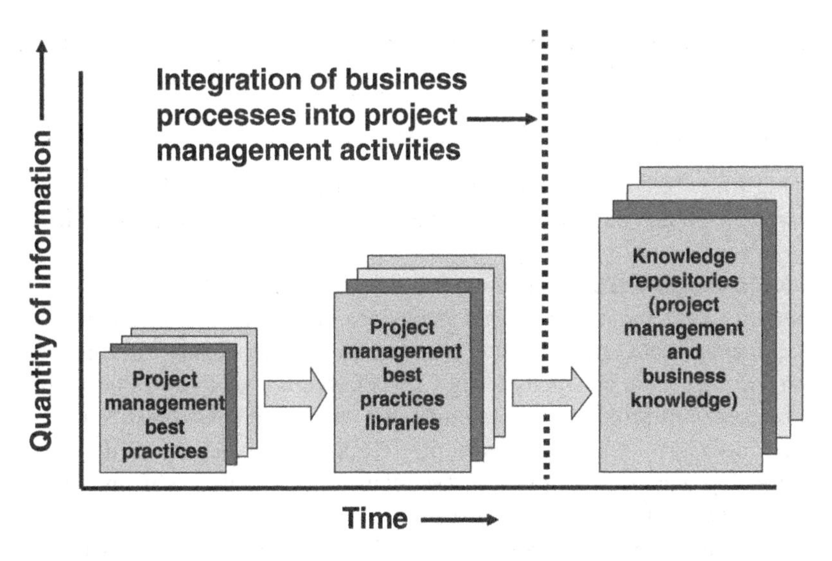

Figure 11.3 The Growth of Knowledge Repositories.

The critical question is, "How often should they be reevaluated?" The answer to this question is based upon how many best practices are in the library. Some companies maintain just a few best practices, whereas large, multinational companies may have thousands of clients and maintain hundreds of best practices in their libraries.

If the company sells products as well as services, then there can be both product-related and process-related best practices in the library in addition to the strategic best practices contained in their information warehouse and knowledge management system.

Every company has its approach to how frequently the reevaluation should take place. Some companies do it quarterly, semiannually, or yearly on the anniversary date of the best practice. Members of the reevaluation committee can change from company to company and may include PMO representatives, best practices owners, subject matter experts, consultants, and the library administrator.

Three types of decisions can be made during the review process:

- Keep the best practice as is until the next review process
- Update the best practice and continue using it until the next review process
- Retire the best practice from service

11.2.8 Utilization of Best Practices

A critical decision that must be made by the guardians of the best practices which is frequently the Portfolio PMO, is deciding what to do with the best practices once discovered. Given that a best practice is an activity that may lead to a sustained competitive advantage, it is no wonder that some companies have been reluctant to share their best practices. Therefore, what should a company do with its best practices?

The most common options available include:

- **Sharing knowledge internally only**: This is accomplished with the use of a best practices library and using the company's intranet to share information with employees.
- **Hidden from all but a select few**: Some companies spend vast amounts of money on the preparation of forms, guidelines, templates, and checklists for project management and view their best practices as proprietary information. These documents are provided to only a select few on a need-to-know basis and may be password protected. An example of a "restricted" best practice might be specialized forms and templates for project approval where information contained within may be company-sensitive financial data or the company's position on profitability and market share.
- **Advertise to the company's customers**: In this approach, companies may develop a best practices brochure to market their achievements and may also maintain an extensive best practices library that is shared with customers after contract award. Some companies identify best practices in their proposals as part of competitive bidding efforts to gain a competitive edge. The use of the best practices may then appear in performance reports. In this case, best practices are viewed as competitive weapons to win future business but run the risk of possibly disclosing proprietary information.

11.2.9 Knowledge Transfer

Knowledge transfer is one of the greatest challenges facing corporations. The larger the corporation, the greater the challenge of knowledge transfer. The situation is further complicated when corporate locations, partners, and contractors are dispersed over several continents. Companies are now using virtual project management teams and are struggling with the implementation of best practices. Without a structured approach for knowledge transfer, corporations can repeat mistakes as well as miss valuable opportunities. Corporate collaboration methods must be developed to maximize the effectiveness of best practices. Effective knowledge transfer techniques can provide significant benefits. According to Georgieva and Allan (2008):

> When the process of knowledge transfer is managed well, knowledge will flow, accumulate, and build up and this will promote better management in the team, the project work, the customer, other stakeholders, and every aspect of project management. Knowledge transfer increases motivation and is an essential element in good leadership.

There is no point in capturing best practices unless the workers know about it. The problem is how to communicate this information to workers, especially in large, multinational companies. The responsibility for knowledge transfers is being placed in the hands of the Portfolio PMOs. Some of the knowledge transfer techniques for best practices that are being used include:

- **Best practices libraries**: This is the most common technique if workers are willing to use the library.
- **Best practices case studies**: Some companies write case studies on their best practices and use them in training programs. The problem occurs when the company has a vast best practices library and cannot prepare case studies for every best practice, and when best practices are updated.
- **Internal seminars**: This technique may provide the most fruitful results. It exposes the audience to the benefits of project management best practices and encourages them to search on their own for beneficial continuous improvement efforts. The speakers are usually some of the best practice owners.

A component of knowledge transfer must include visible support from senior management. Workers must be aware that best practices and continuous improvement efforts will be enforced by all levels of management, not just the PMO.

11.2.10 Sustainment

Companies spend a great deal of time and effort capturing best practices and expect an ROI from their use. But why go through the complex process of capturing best practices if people are not going to use them?

Performing periodic audits to verify the use of best practices is a necessity. This is normally part of the responsibility of the PMO. Although the PMO may have the authority to regularly audit projects to ensure the usage of a best practice, they may not have the authority to enforce the usage. The PMO may need to seek assistance from the head of the PMO, the project sponsor, or various stakeholders for enforcement.

11.2.11 What Does the Next Decade Hold for Best Practices?

With the unprecedented rate of change and the disruptive impact of technologies, the impact of best practices on project management practices is immense.

If we look at the enablers of cloud solutions and artificial intelligence (AI) and their integration with the Internet of Things (IoT), we could strongly shift the dialogue of best practices in the next decade to the human side of best practices highlighted earlier in this white paper. **The future of best practices sits in the hands of people who commit to putting these practices into action toward more sustainable outcomes for their businesses and projects.**

Among the most critical challenges of the next decade is how companies will use their organizational agility to encourage widespread use of best practices. As development and improvement of best practices gain momentum, and with increasing transparency across companies and their partners' ecosystem, there is room for strong co-creation of the next practices that will expedite the pace of maturing the project management profession and its very useful practices.

Benefiting from this increased agility in the way we work in this new decade will make the capture and use of effective practices part of the DNA and mindset of forward-thinking organizations. This includes the ownership of best practices that we highlighted as driven by Portfolio PMOs. Responsibility and ownership of best practices could expand to boardrooms and executive leadership who would be the true course setters for tackling the implementation obstacles we highlighted earlier.

11.2.12 Conclusion

The future could see a much tightly connected ecosystem that sees project management capabilities and practices shape organizational and country agendas for diversity, inclusion, social equity, and other major climate, energy, and infrastructure decisions facing our world. The critical tension about what best practices to focus on and who should have access to these needs to shift to a decision-making dialogue centered on strategic value and associated trade-offs. As we achieve a clearer position for projects as strategic vehicles for transformation and change management, we must ensure that we have the highest ROI on both knowledge capture and knowledge transfer in future organizations.

The recommendations made throughout this white paper point to the increasing importance of the human side of best practices. Organizations that are leading the growth and sustainability agendas of the future will have to create safe cultures where fear of extracting best practices from what has not worked disappears. Acknowledging best practices does not mean that we have reached excellence is also a vital ingredient to the infinite value that continuous improvement must maintain. We can't afford to stop growing and stretching!

As we continue the journey into the next decade and look positively at technological disruptions as enablers for the change in where project teams spend their time, growth of best practices will accelerate. Our lessons learned and retrospections will take on a new meaning and will be closely linked to the strategic agenda of the organization. Portfolio PMOs will continue to mature with wider enterprise impact that responds well to future strategic risks that are embedded in the realities of the VUCA business environment of today and tomorrow.

References

Alam, M., Gale, A., Brown, M., and Khan, A.I. (2010). The importance of human skills in project management professional development. *International Journal of Managing Projects in Business* 3 (3): 495–516.

Anderson, B., Henriksen, B., and Aarseth, W. (2007). Benchmarking of Project Management Office Establishment: Extracting Best Practices. *Journal of Management in Engineering* 97–104.

Chen, E.T. (2015). Emerging Trends in Project Management: Expediting Business, *Proceedings for the Northeast Region Decision Sciences Institute (NEDSI)*. 1–13.

Gartner Group (2000). The Project Office: Teams, Processes, and Tools.

Georgieva, S. and Allan, G. (2008). Best practices in project management through a grounded theory lens. *Electronic Journal of Business Research Methods*. 6 (1): 43–51.

Kerzner, H. (2018). *Project Management Best Practices: Achieving Global Excellence*, 22. Hoboken: John Wiley Publishers.

Tip

In building the dynamic future organization, leaders should invest in spreading the right behaviors, coupled with a dedication to learning and use of best practices.

11.3 The Adapting Value

Values and core values are unique identifiers of what the organization's culture is built on. As we shift to this adaptive future organization, the values chosen need to emphasize what is important around here, and then commit the organization to changing on that path of that transformation. Agile practices have been affecting the ways of working for years and some of the agile principles will continue to dominate the creation of this improved adaptability.

In the following case study, multiple points are made about the agile practices, their impact on the style of leading, the expected ways to inject these practices into the DNA of the organization, and the possible risks associated with the types of changes this requires both internally and in working with customers.

Case Study

Agile (A): Understanding Implementation Risks

In the past decade, many information technology (IT) companies have changed their systems development life cycle methodology to a more flexible framework approach, such as Agile and Scrum. The basis for these flexible frameworks was the Agile Manifesto, introduced in February 2001, which had four values:

1. Individuals and actions over processes and tools
2. Working software over comprehensive documentation
3. Customer collaboration over contract negotiation
4. Responding to change over following a plan

The primary players in the framework included a Scrum Master—a product owner who represented the customer and the development team. The framework allowed for evolving scope and changes to be made while keeping the customer involved during the entire project. Work was broken down into sprints of two to four weeks, in which the team performed the tasks needed to be completed.

Without the formality of a rigid methodology, the need for very detailed up-front planning, and expensive and often unnecessary documentation, projects could be aligned quickly to the customer's business model rather than the contractor's business model. Alignment of the framework to the way that the customer does business, combined with continuous and open communication with the customer, often was seen as the primary driver of project success, an increase in customer satisfaction, the creation of the desired business value the customer wanted, and repeat business.

The increase in the success rate of IT projects did not go unnoticed in other industries. But several questions needed to be addressed before these flexible methodologies could be applied to other types of projects and adopted in other industries.

- Will flexible methodologies work on large, complex projects?
- What if the work cannot be broken down into small sprints?
- What if the customer or business owner will not commit resources to the project?
- What if the customer does not want continuous communication over the life cycle of the project?
- What if the customer wants detailed up-front planning?

- What if the customer says that no scope changes will be authorized after project go-ahead?
- What if the customer will not allow for trade-offs on time, cost, and scope?
- What if the customer uses a rigid methodology that may be inflexible?
- What if your methodology cannot be aligned to the customer's methodology?
- What if during competitive bidding, the customer either does not recognize or understand the Agile/Scrum approach or does not want it used on the project?
- What if the customer wants you to use its methodology on the project?
- What if you are one of several contractors on a project and you all must work together but each contractor has a different methodology?
- Can the use of a flexible methodology or framework prevent you from bidding on some contracts?
- Can your company maintain more than one methodology throughout the company based on the type of project being undertaken?

Remco's Challenge

The executive staff at Remco Corporation was quite pleased with the one-day training program they attended on the benefits of using Agile and Scrum on some of their projects. Remco provided products and services to both public and private sector clients, almost all of it through competitive bidding. IT was not required for any of the products and services Remco provided. Agile and Scrum had proven to be successful on internal IT projects, but there were some concerns as to whether the same approach could be used on non-IT-related projects for clients. There was also some concern as to whether clients would buy into the Agile and Scrum approach.

Remco recognized the growth and acceptance of Agile and Scrum as well as the fact that it might eventually impact its core business rather than just internal IT. At the request of Remco's IT organization, a one-day training program was conducted just for the senior levels of management to introduce them to the benefits of using Agile and Scrum and how their techniques could be applied elsewhere in the organization. The executives left the seminar feeling good about what they heard, but there was still some concern as to how this would be implemented across possibly the entire organization and what the risks were. There was also some concern as to how their clients might react and the impact this could have on how Remco does business.

(Continued)

Case Study (Continued)

The Need for Flexibility

Remco was like most other companies when it first recognized the need for project management for its products and services. Because executives were afraid that project managers would begin making decisions that were reserved for the executive levels of management, a rigid project management methodology was developed based on eight life cycle phases:

1. Preliminary planning
2. Detail planning
3. Prototype development
4. Prototype testing
5. Production
6. Final testing and validation
7. Installation
8. Contractual closure

The methodology provided executives with standardization and control over how work would take place. It also created an abundance of paperwork.

As project management matured, executives gained more trust in project managers. Project managers were given the freedom to use only those parts of the standard methodology that were necessary. As an example, if the methodology required that a risk management plan be developed, the project manager could decide that the plan was unnecessary since this project was a very low risk. Project managers now had some degree of freedom, and the rigid methodology was slowly becoming a flexible methodology that could be easily adapted to a customer's business model.

Even with this added flexibility, there were still limits as to how much freedom would be placed in the hands of the project team. As shown in Figure 11.4, the amount of overlap between Remco's methodology, the typical client's methodology, and the Agile methodology was small. Remco realized that, if it used an agile project management approach, the overlap could increase significantly and lead to more business. But again, there are risks, and more trust would need to be given to the project teams.

The Importance of Value

Remco's project management community had spent quite a bit of time trying to convince senior management that the success of a project cannot be measured solely by meeting the triple constraint of time, cost, and scope.

Rather, they argued that the true definition of success is when business value is created for the client, hopefully within the imposed constraints, and the client recognizes the value. Effective client–contractor communication, as identified in the Agile Manifesto, could make this happen.

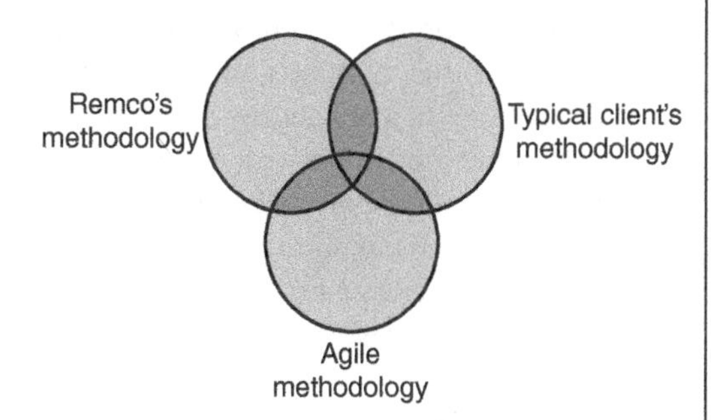

Figure 11.4 Overlapping of Methodologies.

The course reinforced the project managers' belief that measuring and understanding project value was important. Agile and Scrum were shown to be some of the best techniques to use when the value of the project's deliverables was of critical importance to the customers, whether they are internal or external customers. Remco's project management experience was heavily based on the traditional waterfall approach where each phase of a project must be completed before the next phase begins.

This creates a problem with measuring value as shown in Figure 11.5. With the traditional waterfall approach, value is measured primarily at the end of the project. From a risk perspective, there exists a great deal of risk with this approach because there is no guarantee that the desired value will be there at the end of the project when value is measured, and there may have been no early warning indicators as to whether the desired value would be achieved.

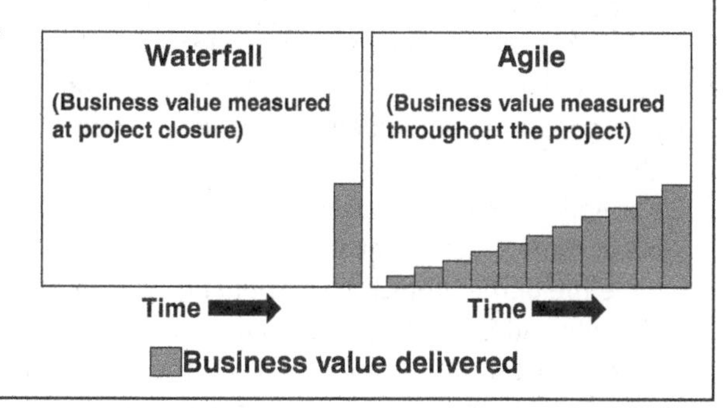

Figure 11.5 Creation of Business Value.

(Continued)

Case Study (Continued)

With an agile approach, value is created in small increments as the project progresses, and the risk of not meeting the final business value desired is greatly reduced. This incremental approach also reduces the amount of time needed at the end of the project for testing and validation. When used on fast-changing projects, Agile and Scrum methodologies often are considered to have built-in risk management functions.

Though other methods of risk mitigation are still necessary, this additional benefit of risk mitigation was one of the driving forces for convincing executives to consider changing to an agile approach for managing projects. Agile and Scrum were seen as excellent ways to overcome the traditional risks of schedule slippages, cost overruns, and scope creep.

Customer Involvement

For decades, project management training programs for public and private sectors recommended that customers should be kept as far away from the project as possible to avoid customer meddling. For this reason, most customers took a somewhat passive role because active involvement in the project could limit career advancement opportunities if the project were to fail.

Remco has some client involvement in projects for the private sector but virtually no involvement from public-sector government agencies. Public-sector organizations wanted to see extensive planning documentation as part of the competitive bidding process, an occasional status report during the life cycle of the project, and then the final deliverables. Any and all problems during the execution of the project were the responsibility of Remco for resolution with little or no client input.

Remco understood that one size does not fit all and that Agile and Scrum might not be applicable to larger projects, where the traditional approach to project management using the Waterfall methodology may be better. There would need to be an understanding as to when the Waterfall approach was better than Agile.

For Agile and Scrum to work on some smaller projects, there would have to be total commitment from the customer and their management team throughout the life of the project. This would be difficult for organizations unfamiliar with Agile and Scrum because it requires the customer to commit a dedicated resource for the life of the project. If the customer does not recognize the benefits of this, then it may be perceived as an additional expense incurred by awarding Remco the contract. This could have a serious impact on competitive bidding and procurement activities and make it difficult or impossible for Remco to win government contracts.

Scope Changes

With traditional project management, accompanied by well-defined requirements and a detailed project plan, scope changes were handled through the use of a change control board (CCB). It was anticipated that, with the amount of time and money spent initiating and planning the project, scope changes would be at a minimum. Unfortunately, this was not the case with most projects, especially large and complex ones. When scope change requests were initiated, it became a costly and time-consuming endeavor for the CCB to meet and write reports for each change request, even if the change request was not approved.

With Agile and Scrum, accompanied by active customer involvement, frequent and cheap scope changes could be made, especially for projects with evolving requirements. The changes could be made in a timely manner without a serious impact on downstream work and with the ability to still provide the client with the desired business value.

Status Reporting

All projects, whether Agile or Waterfall, go through the Project Management Institute's domain areas of initiation, planning, execution, monitoring and control, and closure. But the amount of time and effort expended in each domain area, and how frequently some parts can be repeated, can change. To make matters worse, government agencies often mandate standardized reporting documents that must be completed. Many of these are similar to Gantt charts and other scheduling techniques that take time to complete. Customers may not be pleased if they are told that the status now appears on a Scrum board along with stories.

Government agencies tend to use standardized contracting models, and stating in a proposal that the contractor will be using an Agile or Scrum approach may violate their procurement policies and make Remco nonresponsive to the proposal's statement of work.

Meetings

One of the concerns that Remco's executives had was the number of meetings needed for Agile and Scrum and the number of participants in attendance.

(Continued)

Case Study (Continued)

The time spent in meetings by the product owner and the Scrum Master, and in many cases the team itself, was seen as potentially unproductive hours that were increasing the overhead of the project. With the waterfall approach, meetings almost always resulted in numerous action items that often required months and additional meetings to resolve.

In Agile and Scrum, action items were kept to a minimum and resolved quickly because the people on the team had the authority to make decisions and implement change. This also made it easier to create business value deliverables in a timely manner. There are also techniques available to minimize the time spent in meetings, such as creating an agenda and providing guidelines for how the meetings will be run.

Agile and Scrum work with self-governed teams made up of people with different backgrounds, beliefs, and work habits. Without a definitive leader in these meetings, there exists the opportunity for conflicts and poor decision-making. Without effective training whereby each team understands that they are working together toward a common goal, chaos can reign. People must believe that group decisions made by the team are better than the individual decisions typical of the waterfall approach.

Decision-making becomes easier when people have not only technical competence but also an understanding of the entire project. Effective meetings inform team members early on that certain constraints may not be met, thus allowing them sufficient time to react. This requirement for more information may require significantly more metrics than are used in waterfall approaches. Sometimes executives may be invited to attend these meetings, especially if they have information surrounding enterprise environmental factors that may have an impact on the project.

Project Headcount

In the waterfall approach, an exorbitant amount of time is spent in planning with the belief that a fully detailed plan must be prepared at project initiation and will be followed exactly and that a minimum number of resources will be required during project execution. Risk and unpredictability are then handled by continuous and costly detailed replanning and numerous meetings involving people who may understand very little about the project, thus requiring a catch-up time.

In the waterfall approach, especially during competitive bidding, the client may ask for backup or supporting data as to why project personnel are needed full-time rather than part-time. Some government agencies argue that too many full-time people are an over-management cost on the project.

With Agile and Scrum teams, the scope of the project evolves as the project progresses, and planning is done continuously in small intervals. The success of this approach is based on the use of full-time people who are under no pressure from other projects competing for their services. The people on the project are often rotated through various project assignments; therefore, project knowledge is not in the hands of just a few. The team therefore can be self-directed, with the knowledge and authority to make most decisions with little input from external resources (unless, of course, critical issues arise). The result is rapid feedback of information, a capturing of best practices and lessons learned, and rapid decision-making. Collaborative decision-making involving stakeholders with diverse backgrounds is a strength of the agile approach. Once again, such an approach could be a procurement detriment if the client does not have knowledge of Agile and/or Scrum during competitive bidding activities.

Remco now seemed aware of many of the critical issues and had to decide about converting over to an agile approach. It would not be easy.

Questions

1. Given the issues in the case that Remco is facing, where should Remco begin?

2. What should Remco do if the customers will not commit resources to Agile or Scrum projects?

3. How should Remco handle employee career development when employees realize that there are no formal positions on Agile/Scrum projects and titles may be meaningless? What if employees feel that being assigned to an Agile/Scrum team is not a career advancement opportunity?

4. How harmful might it be to an Agile team if workers with critical skills are either reassigned to higher-priority projects or are asked to work on more than one project at a time?

5. How should you handle a situation where one employee will not follow the Agile approach?

(Continued)

Case Study (Continued)

6 In meetings when there is no leader, how do you resolve personality issues that result in constant conflicts?

7 Can an Agile methodology adapt to change faster than a Waterfall methodology?

8 Will the concept of self-organized teams require Remco to treat conversion as a cultural challenge?

9 Can part of the company use Agile and another part of the company use Waterfall?

Tip

Creating an adaptive value requires fluid experimenting. Agile practices enable us to test many of the adjustments that would complement the required organizational shifts.

11.4 Balancing Steadiness with Change

Ayman Badr, WHO, Head of PMO continues his points by addressing the critical need for balance across projects' portfolios.

> I think based on my experience; we struggle past the mobilization of what we are working on, with:
>
> - The challenge is not necessarily in the starting of projects, as everyone has got good project ideas. Projects start all the time
> - Portfolio excellence is about making sure this project leaders succeed in stopping those projects that should be stopped
> - Leadership is hungry for mechanisms and data for enhancing our decision-making abilities, that would allow them to say yes, we're going to start this, yes, we're going to inject some energy in this, yes, we're going to continue this, and yes, we're going to stop this
> - Having the ability to make those first kind of decisions, where stopping a project, might be taboo in some organizations, and celebrating their

stoppages, as this is not a bad thing to stop this project, is a strategic competency
- Elements around effective decision-making, about what the start, what to accelerate what to stop in organizations, that don't typically work in that way, is a fantastic practice.

Creating that aspired steadiness while building adaptability requires strategic minds that take the time to think of the trade-offs, assess readiness for change, and implement a clear roadmap that suits the movement to adaptability at the right pace. In the following case study, the case highlights the mindset shifts necessary on the part of management, the project leader, the business leaders, and ultimately the teams involved in the adaptive ways of working toward an iterative accomplishment of the project scope.

Case Study

Agile (B): Project Management Mindset[1]

Jane had been a project manager for more than 15 years. All of her projects were executed using traditional project management practices. But now she was expected to manage projects using an agile approach rather than the traditional project management approach she was accustomed to. She was beginning to have reservations as to whether she could change how she worked as a project manager. This could have a serious impact on her career.

The Triple Constraints

Jane believed that clear scope definition, sometimes on a microscopic level, had to be fully understood before a project could officially kick off. Sometimes, as much as 30–35 percent of the project's labor dollars would be spent on scope definition and planning the project. Jane deemed the exorbitant amount of money spent planning the project a necessity to minimize downstream scope changes that could alter the cost and schedule baselines.

Senior management was adamant that all of the scopes had to be completed. This meant that, even though senior management had established a target budget and scheduled end date, the project manager could change the time and cost targets based on the detailed scope definition. Time and cost had flexibility in order to meet the scope requirements.

(Continued)

1 Kerzner, 2022/John Wiley & Sons.

Case Study (Continued)

With agile project management, Jane would have to work differently. Senior management was now establishing a budget and a scheduled completion date, neither of which were allowed to change, and management was now asking Jane how much scope she could deliver within the fixed budget and date.

Planning and Scope Changes

Jane was accustomed to planning the entire project in detail. When scope changes were deemed necessary, senior management would more often than not allow the schedule to be extended and let the budget increase. This would now change.

Planning was now just high-level planning at the beginning of the project. The detailed planning was iterative and incremental on a stage-by-stage basis. At the end of each stage, detailed planning just for the next stage would begin. This made it quite clear to Jane that the expected outcome of the project would be an evolving solution.

Command and Control

Over Jane's 15-year career, as she became more knowledgeable in project management, she became more of a doer than a pure manager. She would actively participate in the planning process and provide constant direction to her team. On some projects, she would perform all of the planning by herself.

With agile project management, Jane would participate in just the high-level planning, and the details would be provided by the team. This meant that Jane no longer had complete command and control and had to work with teams that were empowered to make day-to-day decisions to find the solution needed at the end of each stage. This also impacted project staffing; Jane needed to staff her projects with employees whose functional managers felt they could work well in an empowered environment.

Jane's primary role now would be working closely with the business manager and the client to validate that the solution was evolving. As project manager, Jane would get actively involved with the team only when exceptions happened that could require scope changes resulting in changes to the constraints.

Risk Management

With traditional project management that was reasonably predictable, risk management focused heavily on meeting the triple constraints of time, cost, and scope. But with agile project management, where the budget and schedule were fixed, the most critical risk was the creation of business value. However, since the work was being done iteratively and incrementally, business value was also measured iteratively and incrementally, thus lowering some of the risk on business value.

Questions

1 How easy would it be for Jane to use an agile project management approach from this point forth?

2 If Jane could change, how long would it take?

3 Are there some projects where Jane would still be required to use traditional project management?

4 Empowerment of teams is always an issue. How does Jane know whether the team can be trusted with empowerment?

Tip

Achieving balance and steadiness while shifting to adapting delivery requires that we revisit our view of project constraints and the ways of empowering project teams.

11.5 Co-Created Execution Plans

Adaptable future organizations build on a foundation of co-creation. These organizations listen well to their stakeholders, build on their experiences, and emphasize curiosity and learning as an anchor for realizing effective work outcomes. Co-creation across portfolios of projects is an opportunity to increase the level of buy-in and the assurance that the mix of components chosen widely represents the strategic needs of the organization.

Co-creation is an opportunity to mix diverse points of view, cultures, geographies, and in some cases, extremely opposing positions and ideas. This strengthens

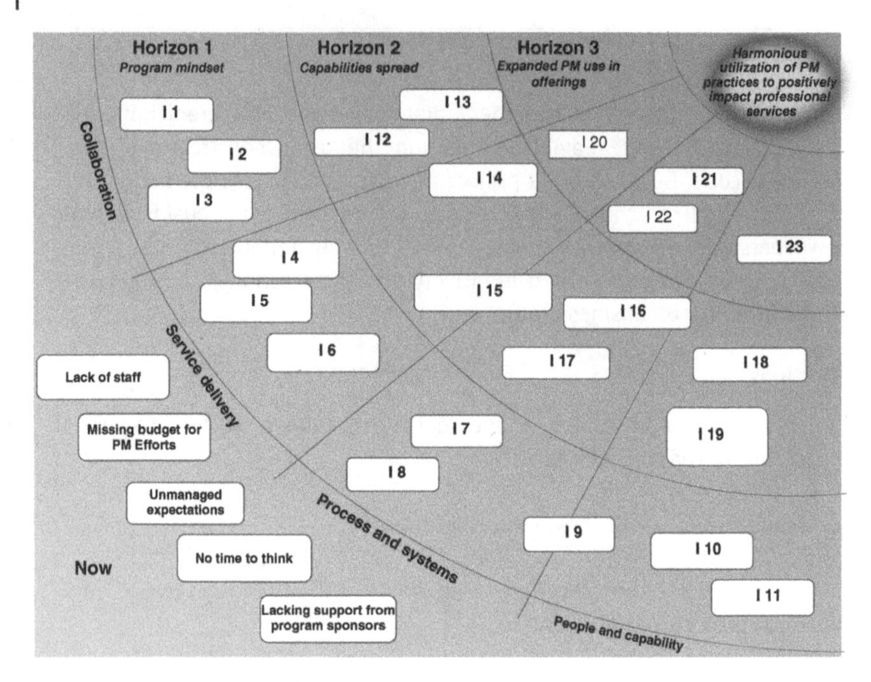

Figure 11.6 The Co-Created Execution Plan.

the collaboration value across teams. Figure 11.6, shows such an example of an execution plan that could be the result of co-creation efforts across key groups of stakeholders.

The visual nature of the plan, coupled with the logical flow of horizons in one of the dimensions, namely horizons 1, 2, and 3, and on the other dimension, the focus areas, such as collaboration, service delivery, process and systems, and people and capability, This structured view of the plan and the transformational nature from the now (current state) into the future state (The NorthStar) along with the right envisioning of the portfolios' initiatives choices, driven by the potential outcomes of these components, gets the organization focused.

The successful use of co-creation also contributes to the success of the next topic of strategic consistency. Consistency is achieved when there is a commitment to ways of thinking and working that would create successful patterns that are replicated across the components of a portfolio of work. When project teams strengthen their co-creation capabilities, they are a step closer to maturing the operational and change fluidity of the organization.

> **Tip**
>
> Co-creating the execution plan, is an opportunity to energize creativity, commitment, and the focus on the transformation path ahead.

11.6 Strategic Consistency

In testing the last one of the eight hypotheses behind this work, the following hypothesis was formulated to investigate the linkages between the right supporting organizational culture and the ability to strategically implement with excellence: "***Building the organizational culture on clear values creates momentum and aligns across organizational portfolios.***"

The key question used for this testing is: *What is most critical for building tomorrow's organizational cultures?*

Figure 11.7, shows the top two scores are: ***clear values*** and a close second being the ***inspiring leadership***. Both topics were previously addressed in this work. Clear values remain critical in connecting the organizational culture and its stakeholders. When the values are clear and simple, this makes the cascading into strategic objectives, success outcomes, transformation initiatives, and ultimately portfolio components and focus, much more realistic. Clear values continue to drive the right behaviors and ultimately connect to inspiring leadership. Inspiring leaders are critical for driving the right results across teams and portfolios of projects.

To create strategic consistency in future organizations, clear values play an instrumental role in achieving this consistency. This strategic consistency matters in defining what strategic success looks like in the future. It ensures that operationally, there is enough level of alignment and learning that the organization can efficiently achieve valuable outcomes repeatedly. This pits the organization on a path of increasing maturity both operationally and in the managing of the portfolio of its transformation initiatives as previously discussed in Chapter 10 of this work.

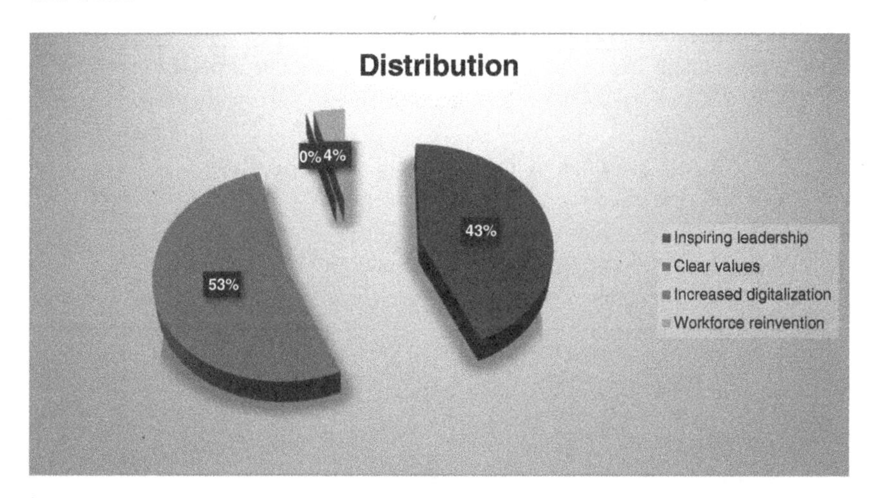

Figure 11.7 The Momentum Creator. *Note*: Based on LinkedIn Open Polling, April 2024.

Tip

Strategic consistently is a key to building the organizational glue for future organizations. Despite the increasing focus on adaptability, leaders need to prioritize this.

Review Questions

Parentheses () are used for Multiple Choice, when one answer is correct. Brackets [] are used for Multiple Answer, when many answers are correct.

1 What is a common definition of best practice?
- ( ) Making sure that there is one way to perform project work better.
- ( ) Providing the desired outcome with more unforeseen complications.
- ( ) A process that is considered superior to other ways of performing the same things.
- ( ) A way to minimize standardization.

2 What are some of the potential benefits of a Portfolio PMO? Choose all that apply.
- [ ] Minimizing the need for risk management.
- [ ] Increasing governance on projects.
- [ ] Projects are aligned to strategic business objectives.
- [ ] Quicker cancellation of failing projects.

3 What is a possible view of project's ambiguity?
- ( ) Interconnectivity of the events.
- ( ) Usually unfavorable changes.
- ( ) Misreading the risk events that can affect the outcome of the project.
- ( ) Events that might occur but being unable to accurately predict.

4 What is a possible value of implementing agile practices?
- ( ) Reduce the need for customer involvement.
- ( ) Usually focus on detailed planning early.
- ( ) Considered to have built-in risk management functions.
- ( ) Changes in scope have serious downstream impacts.

5 What could be key skills for enhancing execution plans co-creation? Choose all that apply.
[ ] Negotiation.
[ ] Multiple languages.
[ ] Siloed thinking.
[ ] Use of diverse expertise.

6 What is a key value of using a clear diagram for the execution plan?
( ) Take more time in developing it.
( ) Increase the level of commitment to the resulting plan.
( ) Focus only on the current state.
( ) Creating a nice picture for the war room.

7 What is a likely key contributor to achieving strategic consistency?
( ) Command and control leadership.
( ) Clear values.
( ) Full digitalization.
( ) High IQ.

Section IV

The Path Forward

Section Overview

"Progress is impossible without change, and those who cannot change their minds cannot change anything." – George Bernard Shaw

The hypotheses behind this work are predicting a future with a high degree of experiencing potential. The ***Experience-Driven Culture*** creates an inspiring environment for the entire ecosystem of stakeholders and technology. The assumption is that in an initiative-based economy, most world organizations will execute their transformation work more strategically in the form of portfolios of projects and programs.

Multiple signs on the path of this organizational maturity were addressed throughout this work as either features, skills, or attributes. Figure S4.1 is like a guiding compass that attempts to capture the most critical elements of this future organizational culture and its associated initiatives' execution capabilities. This compass is built on three layers across the head, heart, and hand principles reviewed earlier.

Layer 1, the ***Foundational Layer***, is tightly connected to the shifts in strategic thinking. It is the *Head Layer* and is likely strengthened by three key building blocks. The first one is the ***New Mindset*** as was emphasized in this work with shifts to taking the time to think, slowing down to go faster, and safeguarding the capacity to experiment.

The ***Innovation Labs*** provides those safe spaces for experimenting and showcase the organizational commitment to shifting how work is executed. The ***Strategic Integration*** building block is centered on the value focus that leaders emphasize in how they formulate strategy and articulate strategic objectives that support value achievement and scaling.

Layer 2, the ***Transformational Layer***, is the *Heart Layer*. This is the secret sauce layer that truly connects the organization and its people around an ethos

Creating Experience-Driven Organizational Culture: How to Drive Transformative Change with Project and Portfolio Management, First Edition. Al Zeitoun.

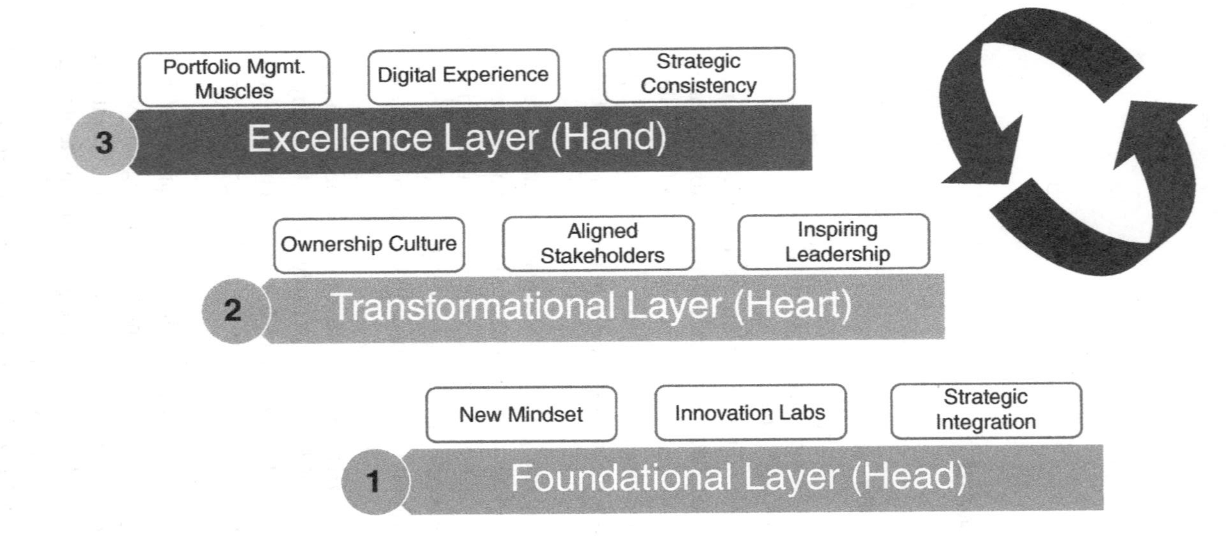

Figure S4.1 Experience-Driven Culture Compass.

and set of values that inspire and motivate movement. The first building block that was highlighted in this work as critical is the creation of the ***Ownership Culture***. This is where the magic happens, when there has been capitalizing on the trust currency, and has enabled the use of critical conversations to ensure commitment and ownership. The second building block, in this layer, is ***Aligned Stakeholders***.

In creating experiencing culture, the most critical ingredient could be how well the organization has managed to build the horizontal working muscle, break down silos, and create a group of highly aligned stakeholders. The third building block is ***Inspiring Leadership***, which was positioned as the ultimate style shift that leaders should do in how they view their role in the future, regardless of the nature of work delivery approach that might be followed.

Layer 3, the ***Excellence Layer***, is where the dynamic execution of the strategic choices and scaling of outcomes take place. It is the *Hand layer*. The first building block is the ***Portfolio Management Muscles***. Maturing the portfolio management practices enables the organization and its initiatives' teams to excel in conducting tradeoffs, refining resource utilization, and focusing the movement toward achieving meaningful outcomes. The second building block is the ***Digital Experience***. In this digital revolution that is expected to dominate the next decade, it is a required shift in experiencing to empower the projects' teams with the power of artificial intelligence (AI), the digital twin, and the potential of the metaverse.

The most effective experiencing could only happen with the power of digital solutions. In order to sustain excellence, humans drive the third building block by focusing on continual balancing and enhancing ***Strategic Consistency***. With the increasing uncertainty, leaders must find ways to use the power of available talent and data to continue the dynamic and iterative strategic adjustments and steady the movement toward a given NorthStar. This is also echoed by the iterative arrows along the compass, as highlighted in the figure.

S4.1 Strategic Opportunities

As summarized by the abovementioned compass, there are many strategic opportunities for setting up and sustaining experience-driven cultures. Organizational readiness will vary across the compass layers, and for some groups, it will be more natural than others to capitalize on the nine building blocks mentioned above. This is also where the tone set by the top leadership is instrumental in moving toward transformation and excellence. These leaders need to have clarity of where the natural strengths of the organization reside and capitalize on those while supplementing the open critical gaps.

In our work, in the PMWJ, Zeitoun and Kerzner (2021), we tackled the importance of being clear on those open critical gaps, and we highlighted a number of the pain points that continue to affect the fluid execution of project work. We also recommend a few concrete steps that the organization could take for the effective cure of some of these organizational viruses.

This paper highlights an integrated review of the pain points that have shaped project management to date and how these pain points created the foundation for the path forward for this profession and the suite of related strategy execution skills. This path forward toward the next decade and beyond will see a continual disruption of the project management principles. Those disruptions are bound to be highly impactful in creating the proper strategic value for project management.

S4.1.1 Background

All too often, business leaders embrace the world of project management and are usually impressed with what they see and hear, especially the benefits that project management can bring to a company along with possibly a sustainable competitive advantage. However, what most people do not see or hear are the pain points that companies have to endure and overcome to achieve their current level of project management maturity and excellence.

Project management pain points began to surface in the latter half of the 1940s, when the U.S. Department of Defense (DoD) invested heavily in the number of projects given out to aerospace and defense contractors following World War II. DoD was the pioneer in developing many of the processes, tools, and techniques that became the foundation elements for today's project management approaches. Project management was also used in the construction industry, although DoD was seen as the primary creator of project management practices.

For many of DoD's contractors, projects brought new types of pain points. Even with the founding of the Project Management Institute (PMI) in 1969 and their publication of the Project Management Body of Knowledge – (PMBOK®) Guide over the years, many pain points persisted. As new approaches to the processes, tools, and techniques for project management appeared, new types of pain points emerged, and these needed to be resolved and mitigated as well.

In this paper, we will discuss key project management pain points most commonly faced by project community today and approaches to tackling them. We will also focus on what we believe are emerging pain points of the twenty-first century and the new multidisciplinary approach for a successful path forward.

S4.1.2 Understanding Pain Points

Historically, pain points have been used most frequently by business analysts or marketers to identify recurring problems, annoyances, or other obstructions that may be inconveniencing their customers. Identifying pain points thereby provides you with the opportunity to sell products or services to customers to relieve the pressure or distress caused by the pain points and position your company as a pain point eliminator.

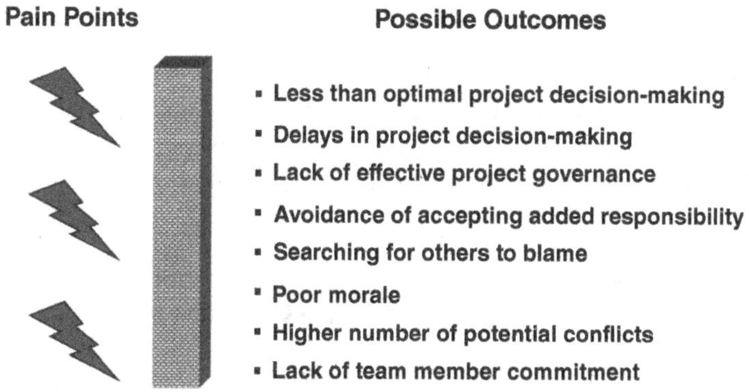

Pain Points **Possible Outcomes**

- **Less than optimal project decision-making**
- **Delays in project decision-making**
- **Lack of effective project governance**
- **Avoidance of accepting added responsibility**
- **Searching for others to blame**
- **Poor morale**
- **Higher number of potential conflicts**
- **Lack of team member commitment**

Figure S4.2 Pain points emanating from all forms of management and leadership activities can impact project management with negative outcomes.

Today, pain points are also being identified in the way that contractors perform the processes needed to satisfy both their company's business model and deliverables expected by their clients. Pain points can be identified in all forms of management and leadership activities, including project and program management, and can create the outcomes shown in Figure S4.2. Pain points can create brick walls that impede successful project management practices and can lead to project failure if not mitigated.

The challenge is in the identification and agreement that some repetitive occurrence is a pain point. What one person perceives as a pain point, another individual may not see it in the same light or see it as an issue that needs to be addressed.

Pain points may appear as simple problems, but a deeper analysis requires establishing pain point categories. In a project management environment with emphasis on the processes, tools, and techniques, typical **Pain Point Categories** might include:

Project Management Pain Point Categories	
Display pain points	Determining the best mixture of metrics and key performance indicators (KPIs) necessary to provide a true meaning of the project's status
Budgetary pain points	Determining the best way to predict the expected cost of the project and potential scope changes that may occur
Scheduling pain points	The inability to eliminate waste and unproductive time from the schedules
Governance pain points	The lack of a structured help line or decision-making process for the timely identification and resolution of critical issues
Methodology pain points	The belief that all projects can fit into a one-size-fits-all methodology

S4.1.3 Traditional Pain Points

Many pain points have persisted over the years. Let's take a look at the most common ones and how companies are tackling them. They cover themes related to communications, organizational politics, the critical role of proper project sponsorship, challenges around career path and standardization, and the potential of project management.

S4.1.4 Customer Communications as Seen by the Contractor or a Third Party

During most of the 1900s, corporate strategic planning was built around the product-market element, namely products offered, and markets served. This implied that the marketing and sales organizations were the dominant players in the formulation of a strategic plan. Most companies appeared to be sales- or marketing-driven because marketing and sales functions were seen as being the primary revenue generators.

Salespeople believed that they "owned" the customers and should be the primary communications link with customers, even though many companies assigned project managers (PMs) to the activities to support salespeople. PMs communicated with the sales team, and then the sales team relayed the information to the customers.

Senior management, who also participated in the communication processes, allowed the communications process to happen. The relationship between the sales personnel and the customers was seen as a strategic necessity, and almost everyone dismissed the fact that it could also become a serious pain point.

S4.1.5 Customer Communications as Seen by DoD

As the number of contracts began to increase, DoD recognized that talking primarily with the sales force about project issues was time-consuming and not necessarily productive. Whenever DoD had technical questions that needed answers, sales would eventually get back to DoD with answers. This process often took weeks for an effective response to the customers.

DoD wanted to talk directly with the PMs and the technical people who could provide immediate answers to their technical questions. The sales force did everything possible to prevent this from happening because they were afraid that they would then have to share year-end bonuses with PMs.

Management persisted in supporting the sales force as the primary communications link to their valued customers. DoD then made the decision to invoke the Golden Rule by holding up the government checkbook to the eyes of senior management and stating, "He who has the gold, rules!" Senior management now felt the pain of possibly having to restructure the company to satisfy the needs of their customers or lose business.

S4.1.6 Project Management Becomes a Career Path Position

In the early years of project management, most companies did not treat project management as a career path position. Project management was seen as a part-time position to be filled on a temporary basis while performing one's normal functional responsibilities. PMs had their performance reviews conducted by their functional managers, and their performance evaluations were often heavily based upon their overall service to their functional unit, rather than based on the success or failure of the project they were managing on a part-time basis.

Government agencies began insisting that project management become a career path position. This created a serious pain point for senior management. Traditionally, contractors were awarded lucrative contracts through competitive bidding. Technical personnel, and sometimes the PMs, would provide input to the sales force who had the responsibility for preparing and submitting the final competitive bid to the client.

The relationship between the sales force and PMs was becoming tenuous. In 1970, an aerospace and defense contractor made the decision to fire everyone in marketing and sales except for one marketer. PMs were then asked to write the proposals and sell them to clients. When asked what the primary skills were for selecting someone to fill a PM position, an executive commented, "communications and effective writing skills."

The government's pressure of invoking the Golden Rule had forced senior management into making project management a career path position. But the pain point was still there. Because many of the contractors were in the aerospace and defense industry, most of the PMs were engineers with advanced degrees in some technical field, with less-than-competitive customer communication and writing skills. Aerospace and defense contractors created technical writing departments to assist the PMs with proposal preparation.

Making project management a career path position also brought with it the pain of having PMs who had never taken any courses in interpersonal skills training or effective leadership. On short-term projects, management endured the pain and instructed the PMs to remove people from the project as quickly as possible after the team members completed their job to keep the project costs as low as possible. It was not uncommon for the PMs to have very little contact with the team members and to rely heavily upon functional managers to provide the necessary day-to-day leadership and direction to their functional employees assigned to project teams.

On long-term projects or those that may have behavioral issues, several aerospace companies assigned organizational development (OD) specialists to assist the PMs. Several years ago, one of these OD specialists was a student in one of the author's graduate courses in project management. When asked what he/she role was on projects, he/she stated that he/she job was to help the PM resolve conflicts and other behavioral issues. He/she also stated that he/she knew very little about the technology on any of the projects and that he/she role was mainly mitigation of behavioral issues.

Allowing engineers with advanced degrees to manage long-term high-technology projects brought with it additional pain points. Some highly technical PMs viewed their projects as a chance for fame and notoriety. As such, their goal was to exceed the specifications rather than simply meet them, regardless of the cost overrun. Military personnel that provided the funding for the cost overruns knew this was happening and believed that, after their 2–3-year tour of duty was completed on this assignment, they would be transferred to another assignment shortly, and their replacement would then have to deal with the cost overruns which were often greater than 300–400%.

S4.1.7 Project Sponsorship

Making project management a career path position was certainly recognized as a pain point that management knew would happen. However, there was an accompanying pain point that needed to be addressed quickly, namely the chance that the PMs would make some decisions that were reserved for the senior levels of management. How could senior management control or influence the decisions made by the PMs, including those decisions that have a serious impact on the business or may lead to unwanted cost overruns?

The answer was in project governance, by providing a project sponsor to oversee and participate in project decision-making. The pain point was mitigated by assigning project sponsorship positions to all the critical projects and staffing the positions with senior- and middle-level managers.

As the number of projects increased, senior managers tried to decrease the number of projects they personally sponsored because it became a time-consuming effort that distracted them from their other duties. Unfortunately, this brought to the surface another pain point on customer (specifically DoD) interfacing.

Many of the military officers that controlled the funding for the projects did not consider the PMs and even some of the sponsors from middle-level management and lower as being equal to them in rank and status. The motto was "Rank Has Its

Privileges," and government personnel persisted in wanting to communicate only to the senior level of management believing that these people were equal to them in rank. As such, senior management was forced to remain as sponsors on many critical projects.

Another related pain point to sponsorship was the impact that a failed project could have on the sponsor's career. If an executive believed that having their name attached to a project that could potentially fail would damage their career path opportunities, they would assign people beneath them in rank as sponsors. If they were still forced to remain as sponsors, they would create a plan whereby others could be blamed if a failure occurred.

As an example, two executives in a telecom company acted as sponsors on two "pet" projects to create new products that they believed would increase sales and generate larger executive bonuses. Both projects required innovation and were costly endeavors. At each of the project review meetings, the PMs recommended canceling the projects because of the significant costs of developing two products that might not generate the expected revenue streams. Both executives (the project sponsors) believed that canceling the projects could impact their careers because of the funds expended. Therefore, at each project review meeting, they allowed the projects to continue to the next gate review meeting. Eventually, both projects were completed at significant cost overruns.

To avoid the embarrassment of having to explain what happened when the marketplace was not interested in purchasing these products, the sponsors promoted the PMs for having developed the products but then blamed marketing and sales personnel for not finding customers for the new products. This reduced and even eliminated the sponsorship pain points.

S4.1.8 Standardization of Processes

As the number of funded projects increased, DoD realized that effective control over the continuously increasing number of projects was troublesome. Each contractor had their own way of performing the work and their own reporting systems. DoD then had the painful experience of having to interpret the data in each status report.

DoD's solution was to establish an **Earned Value** Management **System (EVMS)** and standardize **status reporting**. DoD developed a series of publications encouraging contractors to use the EVMS and government-related life cycle phases and major milestones. Contractors initially saw this new process as a pain point, but soon realized its potential benefits.

Companies often create a one-size-fits-all methodology to be used on all projects. This provided the necessary standardization that executives wanted and made sponsorship and control easier. Performance reviews of PMs and team members were heavily based on how well they followed and used the forms, guidelines, checklists, and templates associated with the one-size-fits-all approach rather than the success or failure of the project.

S4.1.9 Finding Other Applications for Project Management

From the 1970s to the turn of the century, many books and articles cited the benefits of correctly implementing project management and the successes achieved. However, what most people failed to realize was the type of projects that were analyzed to make this determination.

Historically, project management was used on traditional or operational projects that were initiated with a well-defined statement of work (SOW) and work breakdown structure (WBS). Projects for government agencies and most customers were initiated with well-defined requirements. PMs were taught that they should refrain from planning, scheduling, and pricing out a project unless the requirements were very well defined, and techniques were published for finding ways to improve the requirements definition processes. The use of the EVMS worked well on the traditional or operational projects that possessed clear requirements.

But what about the strategic projects, such as those involving innovation activities that are initiated based on an idea rather than rigid requirements, and therefore subject to possible continuous changes? As stated previously, executives were afraid that PM might make decisions that should be reserved for senior management. With well-defined requirements and the use of project sponsors, these risks were minimized. Therefore, PMs would be allowed to manage the traditional or operational projects, while functional managers, whom executives tend to trust more than PMs, would manage the strategic projects.

Giving functional managers control of most of the strategic projects seemed like a good idea, but it eventually brought to the surface a serious pain point that was hidden from view. Functional managers in many companies received year-end bonuses based on the success of the company or their functional unit over the past twelve months. As such, functional managers were retaining their best resources for the short-term projects that affected their bonuses, and the long-term strategic projects were suffering.

There are several reasons why this pain point has been hidden for years. First, functional managers were given the freedom to use whatever processes and techniques they wished to use on their projects. Executives tend to trust their functional managers, allowing them to possibly alter the true status of some of their projects in their reporting to senior management. Second, even if the functional managers used the EVMS, which they mostly avoided, the reporting was on time, within cost and scope, and no information was provided on the quality or capabilities of the assigned resources. Forecast reports, using KPI information, were often an exaggeration rather than reality.

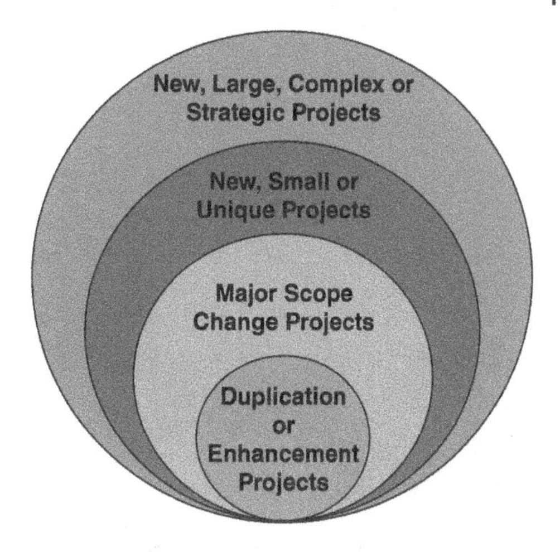

Figure S4.3 New types of projects are emerging, requiring us to think of new approaches of doing and measuring.

S4.1.10 Twenty-First Century Pain Points Appear

The pain points discussed previously appeared, for the most part, in the past. Simply because we have been using project management practices for decades and there have been many successful continuous improvement efforts in project management, does not mean that, in the future, the same or new pain points will not occur.

By the turn of the century, there was a significant growth in the number and types of projects that companies needed to implement for a sustainable and successful future, as seen in Figure S4.3. The greatest change was in the growth of strategic projects as companies realized that continuing to conduct business the same old way was an invitation for disaster.

As the importance of strategic projects became apparent, executives began to ask two questions:

- Do we manage strategic projects the same way we manage traditional projects?
- Do we use the same people, processes, tools, and techniques?

Because strategic projects were significantly more complex than traditional projects, other approaches such as using flexible Agile and Scrum methodologies became obvious. Strategic projects worked better using flexible methodologies and required different skill set training for PMs.

What are some pain points associated with strategic projects?

S4.1.11 The EVMS Becomes a Dinosaur

For more than 50 years, the EVMS has been used with reasonable success supporting the one-size-fits-all methodology, but primarily on traditional or operational

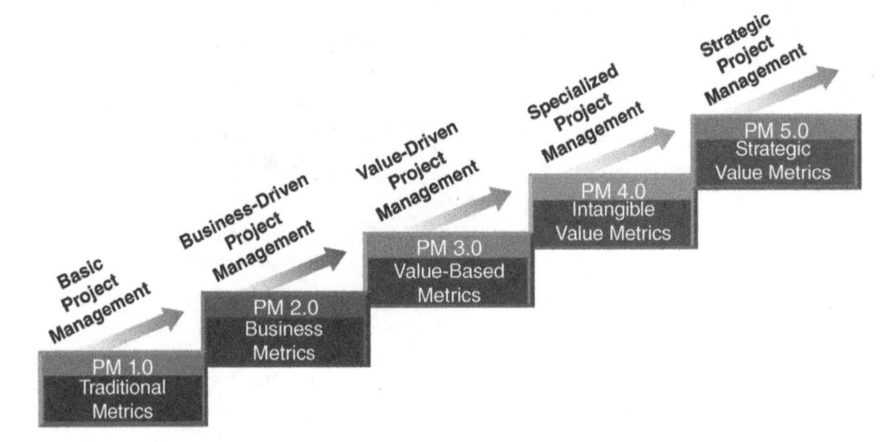

Figure S4.4 The types of metrics used to evaluate project success have evolved from traditional to encompass business and various value-based metrics.

projects. The EVMS focuses mainly on time and cost metrics that can be looked at in various ways to create several approaches for determining project status.

However, as project management practices are being applied to other types of projects, additional information will be required so that management can make decisions based upon facts rather than guesses and intuition. The EVMS system may not become entirely extinct as did dinosaurs, but it is inevitable that it will undergo radical changes.

Strategic projects, where success is measured by business benefits and value created, require significantly more metrics than just time and cost. Companies are now creating their own company-specific project management information systems that may contain as many as 50 or more metrics.

Categories of some of the new metrics are shown in Figure S4.4.

In Figure S4.4, PM 1.0 is traditional or operational project management focusing on just the time, cost, and scope metrics. The business metrics for PM 2.0 were needed because PMs are now seen as managing part of the business rather than just merely a project. These business metrics measure how well project management practices are integrated with the firm's business processes and can measure financial indicators related to the project and market share.

PM 3.0 through PM 5.0 are expansions of business metrics for more detailed purposes. The growth in these three levels is attributed to the fact that today we have become quite good at measurement practices to the point where we believe that we can measure anything.

The following are examples of some of these new metrics being used on projects and can be expected to be components of future EVMS systems and other project management information systems:

- Value-based metrics
  - Competitive position
  - Project's impact on the firm's image and reputation
  - Creation of new business opportunities
  - Possibility for new products and services
  - Impact on speed to market
- Intangible metrics
  - Project leadership effectiveness
  - Project governance effectiveness
  - Functional management's ability to live up to commitments
  - Project morale
  - Customer satisfaction
- Strategic metrics
  - Project's impact on business profitability
  - Health of the project portfolio
  - Business value of the project portfolio
  - Organizational capacity utilization
  - Project's alignment to strategic business objectives

S4.1.12 Executive Support for the New Metrics Management Programs

Sharing information on new metrics, especially business and strategic metrics, requires executive support and discarding the old belief that "information is power." Some executives will not take ownership of a new metrics management system for fear of looking bad in the eyes of their colleagues if the metrics reporting system is not accepted by the workers or fails to provide meaningful results.

Executives tend to not support a metrics management system that looks like pay for performance for executives that can affect their bonuses and chances for promotion. If they were to support such a system, the executives may then select only those metrics that make them look good.

S4.1.13 The Growth of New Flexible Methodologies

The growth and acceptance of Agile and Scrum methodologies have made some people believe that these two flexible methodologies are the "light at the end of the tunnel." It is more likely that these two flexible approaches are the beginning of things to come and a potential pain point executives must face.

By 2030, we can expect to have 20–30 different types of flexible methodologies. The days of having a one-size-fits-all methodology have disappeared. At the beginning of each project, the PM will look at the type of project he/she has

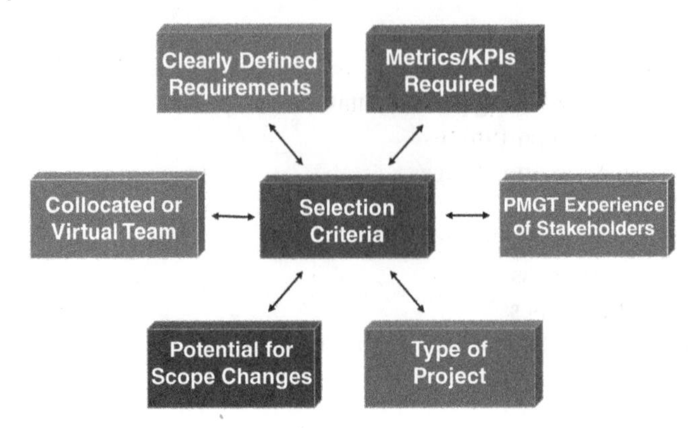

Figure S4.5 Numerous factors influencing the selection of a project methodology.

been asked to manage and then determine the best flexible approach to be taken. A possible list of factors influencing the PM's selection decision appears in Figure S4.5.

S4.1.14 The Path Forward

The multitude of pain points that have been persisting in the field of project management and how organizations have reacted to them have continued into the present state. The highlighted changes reflected in Figures S4.3, S4.4, and S4.5 are positive signs that today's organizations are realizing the potential for the opportunities ahead.

The path toward 2030 and beyond will see the rules of project management re-written so many times, and the pace of those changes is bound to be a higher multiple of anything we have encountered to date.

Figure S4.6 is a brief attempt at predicting some of the future critical dimensions of what we could refer to as the other side of the pain points coin. The idea is to capture what represents the continual tilt we expect to see in how projects are viewed, run, and connect the ways of working for the future organizations.

The other side of the pain points coin is still going to be painful. Strategic shifts of all kinds always are! Let's examine each of these dimensions.

S4.1.15 Culture: Anchoring of New Ways of Running Projects

The cultural changes continue to head in the right direction to support an improved way of conducting project work. These changes tackle several of the pain points highlighted above, especially around connectedness and communications.

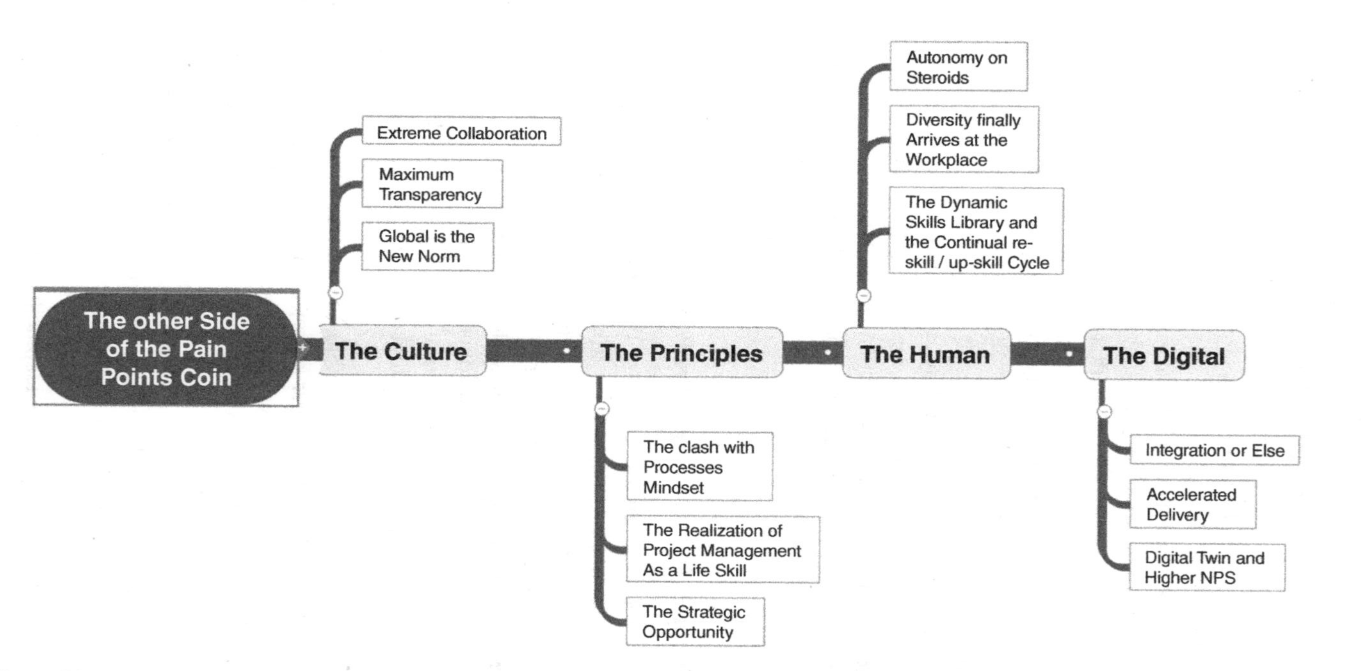

Figure S4.6 The other side of the pain points coin is a continuum of forces involving organizational culture, project principles, human potential, and digital transformation.

Extreme collaboration is the first of three elements of culture. There has been a steady increase in the understanding of the value of collaboration over the last few years, some of which was fueled by many of the organizations forced to conduct their project work virtually and using creative collaboration platforms. This trend has accelerated with the pandemic.

Maximum transparency adds a unique ingredient to project work. With the continual shift to open ways of collaboration, transparency became an expectation. Safety in projects of the future demands that all project team members openly bring tough topics and project concerns to the forefront and early. This openness will finally open the door for real applications of enterprise risk management where risk becomes the dialogue starter for all project team and executives' meetings and discussions.

Global is the new norm as evidenced by the dismantling silos across business silos, disappearing geographic boundaries, and the tightly connected world organizations units. The future organizations will make the study of the research work done around culture maps compulsory. A global outlook will give us the option of realizing the multiple benefits of focused feedback, more aligned decision-making, and better handling of conflicts.

S4.1.16 Principles: The New Fork in the Road

The Clash with Processes Mindset is one of the most promising aspects of how organizations of the future conduct their initiatives. Suddenly, the doors have opened to much more fluidity to the application of project management. Processes will remain a go-to source for organizations that need to maintain some of that rigor, yet the principles will allow them to scale at a higher rate.

The principles could be categorized into buckets such as team, attributes, methods, and success enablers. This dimension of principles has made the PMBOK Guide, 7th edition, a surprising revolutionary change in the practice of the profession.

The shift to principles paved the road to the **full realization of Project Management as a Life Skill**. No longer is project management reserved for the engineers or technical experts of the world. It becomes evident that the project management principles can be practiced by everyone and are no longer a rigid career path.

Strategic **Opportunity** is another value achieved when project principles are practiced widely, easily understood, and clearly communicated. This brings a different take to the role of the sponsor and allows a higher degree of comfort in the type of guidance that is given to PMs and their teams. More effective ways to cascade strategy can be achieved when adaptability and system thinking become dominant. The enhanced understanding of tailoring supports the acceptance by executives of what these project management guiding principles could do to mitigate many of the persisting pains.

S4.1.17 The Human: The Most Stretched Side of the Coin

Autonomy on steroids is an understatement. With the realization that the new generations not only expect but also demand autonomy, workplaces of the future will have to develop all the time. A new Humans 4.0 equation is needed to lead the digital revolution, meet the sense of urgency, and navigate the fluidity of the VUCA (volatility, uncertainty, complexity, and ambiguity) concept that has become the norm. A balancing act with alignment will have to be established as a sanity check for how teams of teams are to tackle future complex programs.

Diversity has firmly arrived at the workplace. Creating inclusive work cultures will increase and be rewarded. We are at an inflection point, where silos can disappear forever. Diversity is the true enabler for the sounds of expertise, the richness of ideas, and the seed of innovation. Inclusivity will be practiced by every leader and refute causes and symptoms of the pain points of the past.

The dynamic skills library and the continual re-skill/up-skill cycle are going to be a feature of the future skills development. Empathy will matter more than ever. Design thinking will rule the day as a center for a multitude of similar ways of working to ensure obsession around customers, and designing mindsets that will focus on the customer experience. This is where continual learning and growing a library of new skills are critical for survival.

S4.1.18 Digital: The Creator of Time to Think Again for a Change

Integration or else is the realization that to stay relevant we will not be talking transformation anymore. Transformation will be business as usual and something that goes on all the time thanks to the digital revolution.

AI coupled with the intelligent Internet of Things (IoT) will give us the foundation to add all the intelligence we need for smarter decision-making. Whether it's trending insights, enhanced estimating, or comprehensive new metrics, digital integration hits a multitude of pain points like a chain reaction. The time that is created for humans to be most valuable via the dedication to thinking, reflecting, and driving more strategically is invaluable.

Accelerated delivery builds nicely on the new flexible methodologies and the rationale mentioned above. With digitally powered experimentation, the benefits of iterative and incremental delivery methods in agile will minimize the cases when we deliver solutions that don't fit or miss achieving stakeholders' satisfaction. Stakeholders truly become the center of the business.

Digital twin and higher net promoter scores (NPS) are a direct result of the new delivery models. When we have advanced simulations and use the richness of data analytics to answer all aspects of a product or a solution, we will be able to go the distance on the lifecycle path of delivering strategic initiatives. This Digital Twin model, which is a virtual model designed to accurately reflect a physical object, allows us to increase the accuracy of the results and the achievement of value. This is also how the NPS, which measures the loyalty of customers to a company, reaches the highest numbers and the scaling impact on the organization accelerates.

This other side of the Project Pain Points coin illustrates deeply that we should not allow perfection to get in the way of greatness. When we combine the enhancements in culture, principles use, develop adaptive and thinking humans, and integrate the enablement value of digitization, we will tackle the pain points and create more focused, healthier, and more connected organizations.

S4.1.19 Conclusion

Reversing the course of the project management pain points has begun. The undivided attention of company executives to the value promise of project management is now a must-have ingredient for extracting the future gains of project management. As we turn the tide on the multitude of pains highlighted in this paper, it is going to take a flavorful sauce to encompass key ingredients such as diversity of transparent inputs, autonomy of working, a human form of digital solutions, and an appreciation of the principles that make the implementation of project management tightly aligned to value.

Comprehending the many points raised above and the history that got us here can become a stretching capability. Learning how we got here allows us to critically focus on where some of our organizational viruses still exist and use that clarity to drive what and how to transform.

The future will have new forms of leadership, and PMs will learn continuously in order to sustain the relevance of projects in making meaningful changes more frequently and consistently. Our hope is to envision a world where boundaries in thought, strategic execution, and results that matter are no more.

Tip

The other side of the pain points coin is the experiencing future. Culture, principles, the human, and the digital come together to shift how we excel in delivering outcomes.

Reflections: In interviewing **Tanya Roberts, IPM, Senior Director, Project Portfolio Management**, she highlights multiple elements of this excellence strategic opportunity.

"I guess the value of culture contributes to what we would call in the consulting world as strategic realization. It's about what are you putting in place to achieve that strategy and culturally, we think there are headwinds and tailwinds. Headwinds are propelling us forward and tailwinds, are sometimes those things that are playing against us.

People have got to be careful of what kind of headwinds are pushing against them. Some of those things are culturally important to the success of an organization. I think first and foremost the leadership and their direction being clear and making sure that the organization is clear on where things are headed, why they're headed that way, and what the measures are that need to support the decision-making process. If decisions have to roll completely to the top culturally because of the kind of organization, this could negatively reflect on the culture.

I think overall communication is culturally critical. People feel part of the company, part of the organization, when the communication is honest, clear, and timely so that they're not the last to know. All of those things combined make for a really important creation of the one culture, not to say that it will be exactly perfect.

If teamwork and collaboration are really strong in your company, then this can propel you forward, especially when you're in enacting change. Your culture will work together to make the change happen. If that's something that they don't do well that can be a headwind, like if we've got silos, or people only get splinters of information, then this could work against them."

Tanya continues to address a few of the excellence attributes:

"I'm assuming that to create joint experiences, it is something to do with allowing open and collaborative and trying new things toward the innovation. I think first and foremost, it is that psychological safety. I think it is really foundational when people feel safe, are able to be part of the culture, to be themselves, to take part in decisions, and have pride in the organization.

From there, it's about feeling empowered to speak one's voice, to be part of making the right decisions, be part of a project team, and really be empowered to do one's part. I think leadership seems to weave its way into many aspects of excellence cultures. Making sure goals are clear, the plan is clear, people are moving in the same direction. Having all of this combined, speaks volumes about how a culture can be very strong. One could follow up the signs of that on the leadership, reflecting what does that look like.

I think it's really important for leaders to have clarity of their strategic direction. Ideally, this is created collaboratively with different leaders, different members of the organization, so then when it is agreed to, rolled out consistently, then everybody feels like they're part of it. In the opposite case, people would feel like the work that they're doing doesn't link to the strategy at all, and then it's easy for them to zone out, not put the effort toward it, because it doesn't mean anything, and they would struggle with how does their work connect."

In addition, Tanya tackled the strengthening of the execution muscle in the organization and the value of the portfolio principles:

"I have been a practitioner of portfolio management for many years and have had the opportunity to work with probably hundreds of companies in my consulting career. There's a lot of portfolio management best practices that companies should follow that they don't.

I'll name a few of what come to mind:

- I think the 1st is that companies need to start stopping saying yes to everything and start saying no to things
- When you lack the focus, a company becomes very diluted, as they're trying to do everything, be everything to everyone, and as a result nothing valuable gets done
- Having the right strategic focus, means saying yes to the most important things
- Really keeping the organization focused I think is the first and foremost
- Secondly, I would say is the best practice of executives needing to do their part, as I I've seen this many times, where executives could be the worst offenders, with their pet projects and wanting to set unreasonable timelines
- I see enhancing governance should be part of the conversation, and this will help mitigate this practice of being the worst offender to the portfolio process
- I think lastly, it is really important to have good a PM foundation, whether it's in the planning or the execution
- I think we forget that that's what makes a portfolio successful is having realistic plans, realistic dates, and thus make great portfolio decisions based on real meaningful data
- When we forget that those fundamentals need to be in place, we have a portfolio that will lose confidence of the organization's stakeholders who will not believe anything that's in it
- Being foundationally very solid, makes a huge difference
- I think with a strong governance in place and an executive team that can understand and respond to the new competing demands, the market changes, strategic goals changes, organizations would be able to relook at the portfolio balance, and whether it still make sense to continue as is
- Sometimes, we move in a different direction, pivot, and make changes, but together as the governance team and not just one person's decision. It's a collective movement toward what's right for the company and not what's right for the individual."

S4.2 Secrets for Leading and Driving

In an interview with **Melanie Winzer, Public Services and Procurement Canada, Director General**, she started tackling a few of these secrets.

"I feel that your culture will determine whether or not you are successful with your strategic objectives. So, whenever I'm developing a strategic plan, or those objectives with my organization, it always centers around a conversation regarding how to enable our people to achieve those objectives within our culture.

For me if we don't have a good culture where employees feel valued, where they can be trained and see progression in their careers, and are mentally safe and have mental health, we would struggle. Wellness and resiliency are incorporated into what leaders do on the day-to-day you won't achieve your strategic objectives without that focus.

For me culture, is one of those key ingredients, if not the main ingredient. You can set up your plans for anything, yet if you don't have the people on your side, you won't get it done. A challenge with culture could be that oftentimes people believe culture is what you grew up in, or what your society that surrounds you is like, or your organizational culture, as well as a blend of all that. If you ask everybody what their definition of culture is, if you ask 20 people, you get 20 different definitions.

I think several attributes are about being open and inclusive. I believe that there are three particular ingredients:

- The first, is diversity because for me it's not just about one voice so I try and pull in people from all walks of life, all ages, all perspectives, and all experiences to build a true team
- We challenge and not do a group think so we don't end up in the wrong place
- The second for me is, I am very much an advocate for mental health and wellness
- I'm a federal speaker where I talk about my own experiences, my lived experiences, so one of the things I do is I build my team and what I feel really feeds into that culture especially when it comes to people is to put that on the table and make sure that everybody feels safe
- Especially when it comes to project management, you need to feel safe to speak truth to power, as a lot of failed projects don't have that component and I've actually had lessons learned that outlined the unsafe work environment was one of the major hurdles for getting to the end zone and it could have been delivered a little bit earlier if they didn't have so much staff turnover
- The third thing for me that I do often, is going through a visioning exercise or something to make sure that we have the same purpose, that we have the same goal and it creates this family like culture
- A joint view of how we can create it, everybody has their place

- I also take a different approach when I know everybody is hierarchical
- Taking an approach to leadership where I've seen myself as that support. I support my team because I can't get it done without them."

Melanie also reflected on some of the execution secrets, with an emphasis on the portfolio management principles' role in balancing investment decisions.

"I really do believe that there are good practices for portfolio management to share across the organization. Over the past two years within my organization, we have created a strategic plan for project management.

One of the biggest things we had to overcome in this journey everybody went through, was that they kept thinking it was my strategic plan for my organization, so I had to tell them that no, this is for our department, for the entire ministry. We are trying to do this together, and that is a big shift to take the time and help create a joint view of the portfolio. What came out of that is a strategic perspective, so one of the things we're targeting now is talking about portfolios, programs of work, and those transformational projects, ensuring that we're not splitting things off because we don't have the money or we don't have the time right now.

Trying to think things through, so that at the end of the day, what we're trying to do, is a main priority that is demonstrating the strategic results. We're trying to achieve results for Canadians social and economic benefits from every single investment, so when you take that lens to look at the portfolio of your projects, it really does change where you invest, how you invest, and when you invest because what you're talking about has to do with strategic choices

- The way we do portfolio tradeoffs, is through the benefits realization approach that I've implemented
- With benefits realization established right at the start, you're clear on what you are trying to do with this investment and the expected outcomes
- You're then very clear on what you're trying to achieve once you develop the cost and the schedule and the scope
- In the past, there has been the tendency if you start running out of money you just would cut some scope or you would cut your schedule or you would extend your schedule. This is not our go to approach anymore, instead what we do is we take a look at the options and say ok this was the outcome we were searching for with this amount of money, this is our return on investment, and this is what we can get out of it
- A second strategic option might be to lower the investment, yet you would still achieve those outcomes just on a smaller scale yeah and then a third option, and then the final option
- If it's too expensive of an investment to achieve those results, then you could cancel

- We've had a couple of those in the last year that we canceled something that was absolutely needed yet the timing just wasn't right, it was just not financially feasible, and/or the schedule was just not possible
- What this does is that it also enables you when risks emerge to realign and determine path forward
- We want to achieve this outcome but we've realized that the assumptions we made were wrong. We now have a choice. We can put more money in or we can leave it as it is and it's just going to take a lot longer so which option do you decide to take?
- In this case, we chose to put more money in, and so it took the project from possibly being delayed by 9 years down to 18 months."

In order to effectively lead and drive the future strategy execution excellence, there are multiple leadership and behavioral shifts that should be central to an organization's agenda toward commitment to excellence. These secrets could be tackled by having a sneak peek into the selection of the next future leader. The secret will be critically supported by an enhanced model of championing change and sponsorship as in the article that will follow the next one.

Two additional enablers will also be highlighted, the first will be demonstrated by showcasing what executive leaders at Booz Allen Hamilton have continually and dynamically shifted in the focus of the organization. The last enabler will be in the analysis of skunk work and how it confirms the many experiencing practices, cultural features, and mindset shifts that were emphasized and highlighted throughout this work.

In our work in the PMWJ, Kerzner and Zeitoun (2024), we tackle the first of the excellence secrets, *the future leader selection*.

S4.2.1 Introduction

Today's literature, whether in textbooks or journals such as PMWJ, abounds with great articles about current and future developments in project management. Sometimes, the articles focus on the mistakes made by PMs and how we can make corrections so that PMs do not repeat the mistakes.

What the literature usually does not discuss are the project management mistakes or actions taken by management that led to problems. In this article, we will look back in time to the early years of project management growth, which is something nice to do occasionally, and look at some of the decisions made by management. Some of these situations still exist today.

Excellence in leading in the future cultures requires us to learn from those past situations and adapt the profile of the next generation of programs and

PMs to respond to the societal demands for the mega change and transformation initiatives that will shape tomorrow's landscape. The level of transparency, intense collaboration, and high degree of innovation and creativity that tomorrow's PMs will have to bring to the table is a critical quality. Engaging leadership matters, and true cultural excellence for delivering value in projects and programs will rest on the shoulders of this new generation of leaders.

S4.2.2 The Journey of Selecting the Project Manager

In the early years of project management, the heaviest users of the discipline were construction companies and the aerospace and defense industry contractors. Almost all of the PMs were engineers, as expected, since the bulk of the projects were technical.

In the aerospace and defense industry, most PMs had advanced degrees. A vice president for engineering in an aerospace and defense contractor was asked who they assign as PMs. He/she responded that the best PMs are engineers with advanced degrees, especially those who also possess good writing skills.

Not all engineers had good writing skills. The aerospace and defense industry contractors solved the problem by creating technical writing departments. Whenever a PM was required to write a technical or status report, a representative from the technical writing department would assist the PM in transforming the words the PM used into expressions that were easily understood and grammatically correct.

While it seemed like the right thing to do, assigning engineers with advanced degrees to the high-technology projects, there were risks. Some of the engineers viewed the assignment as a PM as a chance to increase their image and reputation by trying to exceed the specifications rather than just meeting requirements. The result was often a significant increase in the budget due to scope changes.

This occurred frequently on DoD contracts where the DoD decision-makers were military officers. The military officers also viewed exceeding the specifications as enhancements to their career and knowing that their replacement after their tour of duty ends would be responsible for explaining the rationale for the increase in the government's budget.

Today, just about anyone can have the opportunity to serve as a PM if they are properly educated in project and/or program management. In many cases, some of the best PMs don't come from the classical engineering background.

S4.2.3 Educating Project Managers

Almost all of the government contracts were issued with well-defined SOWs. The problems the PMs had to solve and the decisions to be made were almost always technical. As such, the educational emphasis was on technical training programs. Project management, for the most part, was in the embryonic stage, and companies did not see the need for educating people in interpersonal skills.

The one course that was taught at just about every contractor's organization was the design and use of the EVMS. Government agencies had a multitude of contractors, and each contractor had their own unique way of reporting status. The government was finding it difficult to determine the true status of some of the projects and therefore created the EVMS to standardize status reporting.

Most companies designed and implemented a singular project management methodology built around the EVMS. Educating the organization on EVMS also included instructing the people on how to use the forms, guidelines, templates, and checklists that accompanied the singular methodology.

Future project managers could benefit from finding the most fitting techniques that match tomorrow's projects' demands fluidly.

S4.2.4 Preparing Project Managers

Today, project and program managers are trained using a multitude of educational packages. But years ago, companies believed that PMs should understand how each of the functional units operates and the impact that the PM's decisions could have on each organizational unit. As such, potential PMs would be temporarily assigned to various functional units for a short period of time to understand how each unit operates.

Executives trust project and program managers to make decisions in the best interest of the company and the stakeholders. But in the early years, senior management was fearful that PMs would make decisions in their own best interest and would also make decisions that were reserved for the senior levels of management.

Unwilling to provide PMs with complete trust, senior management assigned project sponsors to every project. The role of the sponsor was to ensure that the PMs were making the correct decisions. Since almost all of the PMs were engineers, the PMs were allowed to make technical decisions, but almost all project business-related decisions had to be made, or at least approved, by the project sponsors that resided at the senior levels of management.

In addition, on almost all government contracts, the contractor's project management team included an assistant PM for contracting and procurement. Sponsors were fearful that the PMs had very little knowledge in these two areas and needed assurance that all decisions and procurement activities abided by legal consideration.

In many companies, the lack of trust ended up creating policies whereby most of the communication with customers and stakeholders was provided by the project sponsors. The sponsors were afraid that the PMs would agree to scope changes that were not funded. In some companies, the situation backfired when sponsors authorized unfunded scope changes to appease the customers and then told the PMs to accomplish the additional work within the original budget.

S4.2.5 Additional Learning Along the Journey

One of the reasons for rotating PMs through various functional departments, even though it was only for a short time, was to provide them with an understanding of how the department functions. The belief was that this would help the PMs develop the best possible project plans.

When many of the project plans failed, senior management made the decision to create a planning department that had the responsibility to produce the project plans for several of the projects. The sponsors usually made the decision as to who would be responsible for project plan development based on the trust they had in the assigned PMs.

In addition to a planning department, several government contractors created a cost control department. The intent was to standardize the project's status in compliance with EVMS. The department served as the "bridge" between the company's singular methodology and compliance with EVMS reports.

Project sponsors attended most of the meetings with the clients and stakeholders and dominated the discussions, even though the PMs were in attendance. Meetings held in the contractor's company were morning meetings. The meetings would adjourn at lunchtime. During lunchtime, the minutes of the meeting were prepared, and the clients and stakeholders were then expected to sign off on the minutes before departing. Today, many of these meetings are now virtual.

Government agencies, especially DoD, often conducted brainstorming sessions for new ideas for products for military use. DoD was unsure what they wanted, and the sessions were conducted before any contract would be awarded.

DoD invited all of the potential aerospace and defense contractors who might bid on the contract to attend the brainstorming session. The expectation was that the contractors would send their best possible technical people to present ideas for future DoD contracts. This would certainly help DoD.

While the intent appeared sound, the contractors most often refused to present their ideas for fear that their competitors could capitalize on the ideas and win future contracts. The result that was the contractors sent some of their best technical people with instructions not to present any critical information but to listen to what others were saying and take notes.

S4.2.6 Stakeholders Project Management Knowledge

In the early years of project management, stakeholders were interested in the results and deliverables rather than understanding how project management was designed to work. Sponsors handled most of the communications with the clients and stakeholders for fear that clients might meddle in the execution of the project and try to change the project's direction without any understanding of how project management was supposed to work. Today, stakeholders are knowledgeable about project management practices, and we welcome their participation.

S4.2.7 The Path Forward

Shifts in how we select future project and program managers are a must. In a future where digitization will change everything we do, the future PM will be the ultimate Empathizer. Investing in selecting PMs who are strong leaders, able to see their initiatives with a holistic lens, and who are capable of making many of the decisions that previously sponsors had to make is the right step toward being an effective PM.

It is vital that tomorrow's PMs are equipped with the proper understanding of the PM practices, yet most valuable would be their ability to remain highly humble and open to the multitude of additional learning and insights that will come their way. The path forward requires investment in PMs who can lead in continual uncertainty and thrive under the endlessly changing conditions. This assumes that we have a new breed of sponsors who know how to get out of the way of the PM and are there to help in connecting some of the business dots, and most importantly, to take any of the bigger rocks out of the way of this next generation of leaders. The Next Gen PMs would need to be prepared to be the Chief Executive Officers (CEOs) of their initiatives as they create the critically needed future transformations.

S4.3 Successful Transformations

The second secret continues to surface as critical for the success of executing transformation initiatives. In our work in the PMWJ, Kerzner and Zeitoun (2022), we dive into the importance of this critical role that enables strategic excellence, namely the champion and/or sponsor for strategic transformation initiatives. Project management educators and practitioners promote the value that effective project sponsorship can bring to projects. Unfortunately, there are many instances where, despite starting out with good intentions, ineffective sponsorship occurs and leads to project disasters and even project failures. In this article, we will discuss several of these ineffective sponsorship situations and what can be done to improve project sponsorship practices going forward.

S4.3.1 Defining Success

For more than 40 years, articles and books have appeared extolling the successes in capturing project and organizational value that can be achieved from the effective implementation of project management practices. While many companies have achieved and maintained high levels of project management success, other companies have limited the continuous investment needed in project management practices to make the success sustainable (Chandler and Thomas 2015; Thomas and Mullaly 2008).

There are many definitions of success in a project management environment. The reason for the disparity is that most companies do not have a clear understanding of the factors that contribute to success (Bryde 2008). For simplicity's sake, project management success and the value it brings to an organization can be described in the following areas: (1) project success, (2) repeatable use of processes, tools, and techniques, (3) impact on the firm's business model, and (4) business results. These areas have been adapted from components in the model used by Thomas and Mullaly (2008).

- Project success has been traditionally defined as completing a deliverable within the triple constraints of time, cost, and scope followed by customer acceptance. The customer could be internal or external to the organization.
- Repeatable use of the project management processes, tools, and techniques is usually a characteristic of success when companies mandate a one-size-fits-all methodology approach for all of their traditional or operational projects.
- Business model success measures the amount of new business generated or an increase in market share because of successful use of project management. It can also measure the effectiveness of portfolio management practices and use of project management offices (PMOs).
- Business results success is usually measured in financial terms obtained from revenue generated from completed projects.

There are other areas of success that could be considered, and many of them are industry-related or dependent upon the type of project. Each of the areas of success can be broken down into critical success factors (CSFs), which are also most often industry-specific.

S4.3.2 Role of the Project Sponsor

One of the CSFs that is common to all areas of success in project management is project sponsorship. Unfortunately, companies have not in the past given sponsorship the attention needed. Companies recognized the need to assign a sponsor, but there was a poor understanding of the role of the sponsor and its importance in delivering project and project management success.

Sponsors were assigned primarily to appease external clients, and executives assigned as sponsors viewed this short and part-time role as an "accidental" or "reluctant" sponsorship assignment. Executive sponsors did not recognize the differences between project sponsorship and executive supervision. The result, as can be seen from the examples that follow, was that ineffective sponsorship soon became a major contributor to project failures, even though sponsors believed that they understood their role and were performing accordingly.

S4.3.3 The Project Sponsor/Project Manager Working Relationship

The birth of project sponsorship began in the early years of project management in the aerospace and defense industries. Most of the projects were highly technical and mandated that engineers, often those with advanced degrees, assume the lead role as PMs. Executive management was fearful that these highly technical PMs would make decisions that were reserved for the senior levels of management, and restrictions had to be in place as to what decisions they were allowed to make.

The mistaken belief was that these PMs, because of their technical expertise, may be ineffective in making project business decisions. This was certainly not true, but senior management preferred to assign sponsors to handle all the business decisions on the projects and let the PMs handle the technical issues.

Many of the people assigned as project sponsors had a poor understanding of project management practices and sometimes the technology as well. As such, the sponsors and the PMs did not communicate as often as needed. The result was that project business decisions were being made without a full understanding of the technology, and technical decisions were being made without an understanding of the impact on the customer and the business.

Result: Poor project decision-making

S4.3.4 Customer Communications

Companies soon realized the abovementioned issue with the relationship between sponsors and PMs. Many companies considered assigning sponsors from the lower or middle levels of management rather than from the senior levels. While this approach was expected to increase collaboration and resolve some of the collaboration issues, government and military personnel did not see this as being in their best interest and exerted their influence.

Many government workers and military personnel believed that, because of their rank or title, their "equals" in the contractors' firms were at the executive levels. As such, even though lower-level individuals were assigned as sponsors, government and military personnel that controlled the funding for the projects communicated only with senior management, thus forcing them to remain as sponsors. Simply stated, sponsorship was often based upon the impact of two government rules:

- Rank has its privileges.
- He/she who controls the "gold," rules! (i.e., makes the final decision).

Senior management succumbed to the pressure and remained as sponsors to appease the customers. Most of the time they functioned as "invisible" sponsors.

Result: Ineffective project sponsorship

S4.3.5 Information is Power

In the early years of project management, senior management believed that allowing PMs to make business decisions was not only a risk but also diminished the role of senior management. Many executives believed then (and some still do) that information is power. Therefore, providing PMs with the necessary strategic or business information needed for business decisions would reduce their power base.

When information is power, project teams do not have a line-of-sight to senior management and therefore make decisions that may not be aligned with strategic business objectives.

Result: Lack of alignment across project teams

S4.3.6 Sponsorship Growth

It did not take long for the benefits of project management to appear. Companies began using project management for internal traditional or operational projects as well as projects for external clients. Now, there was a need for significantly more project sponsors.

Senior management recognized quickly that they could not function as project sponsors for all the projects. Sponsorship could be delegated to the middle or lower levels of management, but they would soon complain about the amount of time they would need to perform as sponsors and the fact that it could force them to reduce their efforts on other activities necessary to support daily activities.

Senior management made the decision that, for the internal projects, the business owners would assume the role of project sponsors. This created additional problems. The business owners had very limited knowledge about how project management should function. Many times, they did not understand the technology or the complexities in developing the technologies or creating product features. But what appeared as the worst situation was when business owners made project decisions based upon short-term profitability that could impact their year-end bonuses and sacrificed the long-term benefits and value the project could bring to the firm.

Result: Short-term project decision-making

S4.3.7 Educating Sponsors

For decades, many of the people assigned as sponsors did not fully understand their role and had very limited knowledge about project management. Some companies set up training programs to educate people on the role of a sponsor. Unfortunately, many of the project owners did not believe they needed to attend such a course, even though most of the courses were less than two hours in duration. They felt that it was beneath their dignity to be told that they must be educated on how to properly function as a project sponsor given the fact that they were all in management positions already. These people believed that project sponsorship was the same as providing executive guidance.

Result: Understating the role of the project sponsor remained unsolved

S4.3.8 The Fear of Becoming a Sponsor

Even with excellent sponsorship, not all projects will succeed. As companies undertook more projects, there were also more failures. The concern that sponsors had was whether the failure of a project under their sponsorship could have a detrimental effect on their careers.

This fear forced some executives and managers to avoid sponsorship entirely or to look for ways of blaming others if a project failed. In one company, two business owners acted as sponsors on two projects they expected would be well accepted in the marketplace. These were "pet" projects that captured the imagination of the sponsors.

At each of the gate review meetings for both projects, the PMs stated that the projects should be canceled because the marketplace acceptance expectations seemed unreachable. Both sponsors were afraid that the cancelation of their projects would be seen as wasting valuable resources and could then detrimentally affect their future ambitions.

Both sponsors made the decision at each gate review meeting to let the project continue to the next gate review meeting for a decision to be made on cancelation in hopes that something good might happen unexpectedly in the meantime. The projects were never canceled. Both PMs completed their projects, but as expected, the marketplace did not appear interested in either product.

To save face, the sponsors promoted both PMs for having completed the deliverables and then blamed marketing and sales personnel for not finding sufficient customers. Marketing and sales personnel became the scapegoats.

Result: Ineffective sponsorship wasted valuable resources

S4.3.9 Sponsor's Role in Project Staffing

PMs are at the mercy of functional managers for qualified staffing for the project. PMs may not know the skill sets needed from the functional groups. But when the PMs do know the skill set and the functional managers provide resources that the PMs consider as inadequate, the PMs naturally expect the sponsors to intervene and assist them in obtaining the correct resources.

Many sponsors have shied away from participating in project staffing for fear of alienating functional managers that they may have to work with in the future. As such, it was not uncommon for sponsors to avoid all responsibilities and participation in project staffing activities where they may have to usurp the authority of other managers.

Project sponsors did not like the idea of telling functional managers in other functional units how to staff a given project, especially since the sponsors did not know what other projects the functional units were responsible for staffing or the accompanying priorities.

Result: Projects are staffed with the wrong resources

S4.3.10 Sponsorship Staffing with a Hidden Agenda

In the previous example, we showed that sponsors may not desire to participate in project staffing. At the other end of the spectrum, we have sponsors who may insist on project staffing participation, especially if the sponsor believes that the success of the project that they are sponsoring could have favorable implications on their career goals. This occurs when sponsors may have a hidden agenda related to this project.

Based upon the sponsor's rank and title, the sponsor may possess the authority to force functional units to staff a project with individuals hand-picked by the project sponsor. This is often done with little regard for the impact of removing the workers from another project that desperately needed their skills.

Result: Project staffing is not done in the best interest of the company

S4.3.11 Making Unrealistic Promises to the Customers

It is not uncommon for sponsors to handle communications with clients, especially with the senior levels of management in the clients' organizations. While this is an acceptable and often beneficial activity, it can create problems when the sponsor makes promises to the client as a way of appeasing the client or simply to look good in the eyes of the client.

As an example, during a discussion with the client, a sponsor promised the client that the company would perform additional testing to validate certain numbers in a report. The cost of the additional testing was more than $100,000. The sponsor told the PM to perform the additional testing within the original budget and that the sponsor would not be pleased if there were any cost overruns. The sponsor wanted this to be treated as a "no-cost scope change."

The project team was unable to hide the costs of the additional work, and the profit of the project was reduced by $100,000. The sponsor reprimanded the team for not following his/her instructions, even though they were unrealistic.

Result: No-cost scope changes rarely exist

S4.3.12 Not Wanting to Hear any Bad News

Sponsors exist to help project teams resolve problems and make the right decisions. Yet there are many sponsors that tell the teams that they do not want to hear any bad news. There are several reasons for this. The sponsor may not want to relay any bad news to the client and feels it is better not to know about the issues. The sponsor may feel that bad news can be detrimental to his/her long-term goals. The sponsor may not wish to be involved in solving problems.

Perhaps the worst case of not wanting to hear bad news was identified as one of the causes for the Space Shuttle Challenger disaster where senior management expected lower-level managers to filter bad news from reaching the senior levels of management. There exists a valid argument that the filtering of bad news led to the death of seven astronauts.

Result: Filtering bad news can create very serious problems

S4.3.13 Lessons Learned

What can be learned from the situations provided here? First and foremost, the success of a project is not entirely under the control of the PM. There can be numerous issues that are outside of the control of the PMs and require involvement and decisions by project sponsors. The role of a sponsor is quite complex and can be different in companies even in the same industry. Without a clear understanding of the roles and responsibilities of a sponsor, it is impossible to determine how sponsorship can and does contribute to project success.

Sponsors need to understand their role and the decisions they are expected to make. They should understand this before functioning as a sponsor, not by trial-and-error when performing as a sponsor. The PM of a telecom company became concerned that her sponsor was making decisions that she was unaware of and often disagreed with. She met with her sponsor. On a whiteboard, she drew a line down the center and listed many of the decisions that she expected would need to be made on the project.

Then she looked at the sponsor and asked for clarification as to which decisions she was authorized to make and which decisions must be made by the sponsor. The result of her meeting with the sponsor was a clarification of the lines of responsibility, which made the organization aware of the issue and eventually led to the creation of a project sponsor's role template that became part of the firm's project management methodology.

S4.3.14 The Need for Sponsorship Standards

Professional organizations have created standards for project management, but there do not appear to be any standards or guidelines for project sponsorship. The UK-based Association of Management (APM) defines a project sponsor as the individual/body, who is the primary risk taker, on whose behalf the project is undertaken, and the US-based PMI describes the sponsor as the person/group that provides the financial resources, in cash or in-kind, for the project. These two definitions characterize a project sponsor as being the primary risk taker or the resource provider (Bryde, 2008).

Companies must understand the CSFs that lead to effective sponsorship and success. The following list, which is not in any specific order of importance, provides some guidance in understanding the role and responsibilities of a sponsor:

- Sponsors must understand that on some projects they may have to function as the primary communications link with the customer.
- PMs may not possess the authority to drive the projects to success without support from sponsors.
- We are now using project management on strategic as well as operational or traditional projects. Sponsors provide the knowledge and authority to make sure that project decisions are aligned with corporate strategy and strategic corporate objectives.
- Sponsors must be assigned during project selection activities to ensure that the best portfolios of projects are chosen and to get their buy-in. Sponsors should possess skills in conducting a Strengths, Weaknesses, Opportunities, and Threats analysis and application of established business models during project selection.
- More and more projects today are impacted by the enterprise environmental factors in the VUCA environment. Sponsors may possess a better understanding than PMs of how the company is impacted by the VUCA environment.
- The VUCA environment increases the risks that the company must face. Sponsors can provide guidance on how to best mitigate the risks.

- Sponsors are more than just business owners who fund projects. They possess the authority to ensure that the correct resources are assigned to the project.
- Sponsors must understand that status reporting is no longer based on just three metrics, namely time, cost, and scope. Sponsors must participate in selecting the proper mix of metrics such that the true project status can be determined quickly and that project sponsorship decisions will be based upon evidence and facts rather than guesses.
- Sponsors must provide PMs with the criteria (perhaps based on metrics selected) as to what will be defined as project success and project failure. Project failure criteria are essential so teams will know when to stop working on a project and wasting resources.
- Sponsors must understand how their decisions impact the outcome of projects and can lead to success or failure.
- Sponsors must recognize that the true success of a project rests in the benefits and value that come from the deliverables. It may be months or years after the project's deliverables have been produced before the real success of the project can be seen. Sponsors must therefore remain active as sponsors over the full life cycle of the project including benefits harvesting and sustainment of benefits and value. This is especially true for projects that lead to organizational changes in the firm's business model.
- PMs rely upon the PMs for guidance, leadership, networking, coaching, and mentorship during the execution of the project. Therefore, sponsors must possess more than just a cursory knowledge of how project management should work.
- Sponsors must be willing to attend periodically sponsorship courses to learn about sponsorship CSFs and best practices.
- Some companies have embarked upon committee sponsorship because one person may not possess all the necessary skills for sponsorship. Other companies have created specialized PMOs that have as their primary function the sponsorship of the portfolio of projects under their control. All members of the sponsorship committee, as well as PMO leadership personnel, must understand the role of a sponsor.

S4.3.15 The Path Forward

Based on the multitude of ineffective leadership results addressed, the gaps in the standards, and in commonly agreed to attributes for effective sponsorship, we recommend a revolutionary investment in maturing two critical dimensions to the success of sponsorship. With projects taking their seat as the strategic vehicles for sustaining the organizations of the future, it has finally become attractive for organizations to seriously consider this sponsorship topic critical investing in their future.

The first path-forward dimension is expanding the mindset around the future role of project and program sponsors. In the future, the main value creator in organizations will be projects. Project management is going to be part of the ongoing dialogue of future executive leaders and thus understanding and investing in the critical attributes of sponsors beyond currently accepted narrow views will be a must. We see possible three critical buckets of attributes that make a difference in the persona of effective future project sponsors.

a. **Project conditions**: PM techniques, common language, extended strategic project life cycle view, metrics objectivity, and impediments removing
b. **Sponsor's character**: Decisiveness, relational capacity, excellence orientation, simplification creator, servant leadership, project value selling, political savviness, and courageous
c. **Sponsor's attitude**: Positiveness, inspirational, visionary, and risk-balanced

The second path-forward dimension pertains to creating a *sponsors' track* in every organization that delivers initiatives. With the project economy growth surpassing every expectation and a much higher focus on achieving unique customer and business experiences, it has become critical that this sponsor role is no longer accidental or left up to the views of the who is in charge at a given moment.

Creating a structured track similar to the simple concept, highlighted in Figure S4.7, could be a good starting point. Ideally, it should be a continuous improvement-based track that combines training, coaching, and practical and maturing sponsorship experiences. We see this sponsor track also as a possible answer to the aspirational growth of many project and program managers who want to expand their career journey to an enterprise-wide level of responsibility. The future organizations are committed to putting effective sponsorship on their strategic radar, and the associated return on investment (ROI) is bound to be worth it.

References

Bryde, D. (2008). Perceptions on the impact of project sponsorship practices on project success. *International Journal of Project Management* 26 (8): 800–809.

Chandler, D.E. and Thomas, J.L. (2015). Does executive sponsorship matter for realizing project management value? *Project Management Journal.* 46 (5): 46–61.

Kerzner, H. and Zeitoun, A. (2022). The connected future business culture: the great project management accelerator; introduction to the series and the maturing project sponsorship. *PM World Journal* XI (I).

Kerzner, H. and Zeitoun, A. (2024). Selecting the next PM, the future project culture. *PM World Journal* XIII (IV).

Thomas, J. and Mullaly, M. (2008). *Researching the Value of Project Management.* Newtown Square, PA: Project Management Institute.

Zeitoun, A. and Kerzner, H. (2021). Project management pain points and a path forward; *PM World Journal*, Vol. X, Issue X.

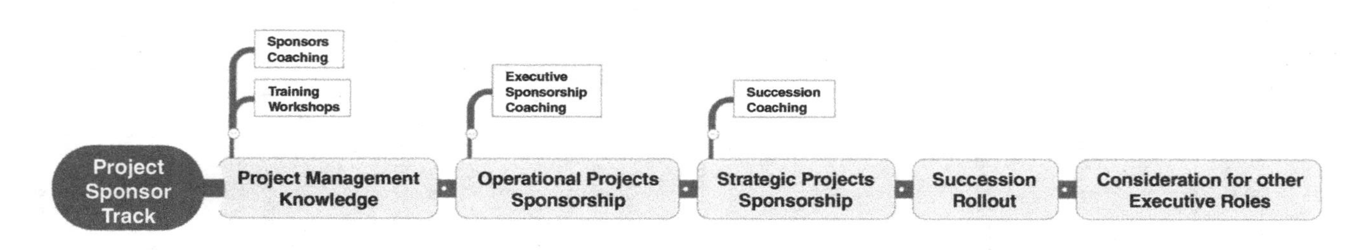

Figure S4.7 Future Concept for a Project Sponsor Track.

> **Tip**
> The mindset shifts in selecting the future leader, coupled with standards and CSFs for identifying the right future sponsors, will unlock key secrets for strategic excellence.

Rob Silverman, Executive Vice President at Booz Allen Hamilton, shares the following insights about how organizational scaling shifts and focused inspiring leadership, sustain the drive for excellence.

S4.4 A Booz Allen Secret Sauce Example

S4.4.1 Introduction

Celebrating 110 years as a company in 2024, Booz Allen Hamilton (Booz Allen) has grown into a Fortune 500 company with over US $9 billion in revenue and over 34,000 employees. In the last decade, Booz Allen has leaned more into Mergers and Acquisitions (M&A)—adding to already-strong capabilities in technical areas of digital software delivery, AI, and cyber; mission and business spaces across public and commercial client spaces; and new and innovative business practices and solutions consistent with our broader corporate strategy.[1]

M&A and its accompanying post-merger integration (PMI) are difficult, a priori, given the requirement to merge cultures into an enhanced value proposition for employees, collectively and individually; to complement and expand brand, client base, and value proposition beyond the sum of its constituent parts; and to achieve proposed deal model synergies. At Booz Allen, we have learned lessons through our recent acquisitions and their integrations with our company, and—while recognizing the uniqueness of each deal, company, and employee base—we've become adept and disciplined about compiling, updating, tailoring, and applying those lessons to our new acquisitions.

Those lessons learned apply to the entire lifecycle of a deal—from pursuing and acquiring a company through its full integration. For the purposes here, we will leave aside deal and business model mechanics and strategy involved with acquisition and instead emphasize program management philosophy and practice in PMI.

---

1 Booz Allen's corporate strategy—named "VoLT" which emphasizes velocity, leadership, and technology—follows from the previous "Vision 2020" strategy, widely acclaimed for delivering stakeholder success while fundamentally transforming Booz Allen from primarily focused on legacy general management consulting offerings to an advanced technical organization with world-class digital, engineering, analytics, and cyber talent and solutions, all underpinned by fundamental consultative DNA.

S4.4.2 Starting with Integration Philosophy and Strategy

At Booz Allen, we start our internal thought process for PMI well before a deal is consummated, including initial meetings between the two parties (Booz Allen and target company) and a retinue of bankers, lawyers, and deal team members. Ultimately, we need to build rapport across organizations—ensuring Booz Allen feels comfortable with the company being acquired and that the company feels comfortable with Booz Allen.

Particularly, in addition to concurrence on enterprise value and associated deal mechanics—saved for a different forum—we align with our proposed new colleagues and Booz Allen integration stakeholders[2] on our shared Purpose and Values, planned points of contact and key relationships post-close, how we plan to tailor our integration approach to the asset being acquired, how we will measure performance, and how we will manage the integration project.

At Booz Allen, we ***empower people****—our colleagues, our clients, and our communities—****to change the world***. We live that purpose through core values of *Ferocious Integrity, Unflinching Courage, Passionate Service, Champion's Heart, and Collective Ingenuity*. Those are not slogans; they are real, and they matter deeply to every individual at Booz Allen. We revel in living through and talking about "real life" examples of how and when colleague(s) demonstrate each. Before we consummate a deal, we talk about our **Purpose and Values** in-depth, highlighting examples, with our target company and ask their colleagues to share their purpose, values, and related experiences. For example, our recent acquisitions have focused on improving care for our Nation's Veterans to keeping our country and our citizens safe and secure at home and abroad to improving the delivery of critical public- and private-sector services. Aligning on our Purpose and Values is binary—if we do not have alignment, we know the deal won't "work" and have withdrawn from consideration when we've detected misalignment.

Early on,[3] we aligned between Booz Allen and our target company on how we would operate as a joint entity. This includes how and to whom the entity will report organizationally and which processes and systems the acquired target will use in the conduct of business (i.e., standard Booz Allen or otherwise).

We've labeled our general PMI approach "***2 DEFT***," and we tailor it to the circumstances of the asset and its employees, our target market(s), and the proposed business model. This includes:

---

2 Key Booz Allen stakeholders include C-suite, client-facing market sector seniors, and enterprise operations leaders responsible for People Services, Finance, Contracts, Legal, Enterprise Technology, Security, Risk/Insurance, etc.

3 Particularly in our primary U.S. Government business space, there are specific rules about how, when, what, and with whom communications can occur before a deal is closed. Our expert Legal and Contracts staff ensure we have, are trained on, and follow proper processes (e.g., on "gun jumping").

- **2-Way Integration:** We approach integrations with the mindset that Booz Allen has as much to learn from the newly acquired company as vice versa—from capabilities to market sensing to efficient and innovative ways to deliver and operate the business.
- **Deliberate Approach:** Documented by our Integrated Master Schedule (IMS), we use a purposeful approach to integration. We resist the urge to do everything at once, often opting for a "crawl, walk, run" approach where we start with mutual goal setting, learning, and incremental victories (e.g., collaborating on single client assignment or research and development (R&D) project) before adding to the list of integrative activities and then ultimately into full, long-term operations.
- **Employee Empowerment:** Employees—both the acquired employees and their Booz Allen counterparts—are at the center of the integration and associated operations. Their experience—from sign through close, integration, and eventual full-scale operations—will determine the success or failure of the integration and proposed synergies. Talent from both organizations should see the merger as an accelerant to the business, to client, and to their career opportunities, collectively and individually.
- **Focus:** Booz Allen's culture of Collective Ingenuity and single profit and loss (P&L) business model enable operational agility, quick reaction to address client needs, and "one team, one fight" approach, but can lead to overwhelming an acquired entity without laser-like focus on specific, prioritized targets of opportunity laid out in the deal's business plan and proposed synergies and continually tracked and updated as necessary. We prioritize and prosecute specific business development, cross-team staffing, and intellectual capital (IC)/R&D building opportunities based on how and how much those opportunities synch with and fulfill our *strategic* intent, the relative *size* of the proposed ROI, and unique *subject matter expertise* (SME) required and/or planned to be built—affectionately called the "3 Ss."
- **Transparency:** We insist on relationships and communications within and across teams that emphasize transparency. Say what you mean, and do what you say. There could be a tendency to continually and overly "sell" the parent or acquired company on the combined entity by avoiding difficult conversations, putting positive spin on difficult situations, or not adequately listening to others' points of view—e.g., if the ultimate strategy requires pursuing a new business over an established but outdated one, how compensation and benefits compare to the acquired asset's previous model, how titles and reporting relationships will work, etc. Thorny issues aren't like fine wine; they don't age well. We use radical candor to communicate directly, with empathy.

We track expected financial performance measures associated with deals as part of routine business operations. Those are part of a more fulsome performance measurement approach that employs the *Balanced Scorecard*. Moreover, we communicate this approach to the target company leaders as early as possible with specific plans and numbers communicated post-close. This not only transparently communicates how we will "keep score" but also how Booz Allen thinks strategically about the business and matches measures to the strategy. For example, in addition to the standard financial measures referenced above, we track customer (satisfaction) measures, employee measures, and internal/other process improvement measures, all around our foundation of business strategy and core values.

Finally, we staff our PMI effort with sufficient representation from the acquired entity and across Booz Allen's market/capability (e.g., leader(s) from the planned client space target(s)) and enterprise operations teams (e.g., People Services, Finance, Contracts, Legal, Enterprise Technology, Security, Risk/Insurance, etc.), occasionally supplemented by specialty contractors skilled in PMI and/or specific functional areas. Our organizational structure includes a full-time Integration Management Office (IMO) that reports to a Steering Committee of C-suite leaders and is supported by matrixed staff from the various functions mentioned above.

We employ a standard WBS with four primary top-line activities, each having 2–5 key sub-activities:

- **Infrastructure Optimization:** How we will merge the acquired asset's enterprise corporate operations (e.g., human resources, information technology, finance, contracts, security, and legal) with Booz Allen's. We emphasize business efficiency—with openness to new ways of doing business—and risk mitigation to seamlessly integrate new colleagues into the Booz Allen family and empower them to continue to operate successfully.
- **Culture Change Management and Communications:** Here is where deals are "won and lost." How does each side—now coming together—learn, appreciate, and capitalize on the other's "secret" or "special sauce"? How do we foster open, efficient communications? Do the acquired and Booz Allen talent bases understand and seek benefit from Booz Allen's stated *employee value proposition* **(Be you. Be Booz Allen. Be empowered)**? Employees who do not see the benefit—from how they are compensated to what growth opportunities they will have—access to new clients, capabilities, assignments, and colleagues; promotions; etc.—will leave, diminishing the enterprise value potential.
- **Client and Market Development:** Merging processes for pursuing, capturing, and prosecuting business to capture deal synergies. Prioritizing those clients, contracts, and partnerships to jointly pursue to make "1 + 1 way more than 2."
- **Capability Development and Deployment:** Starting by learning each other's unique capabilities and IC, determine focused areas for joint IC development—new IC or newly added IC—and delivery to clients. In addition

to differentiating the power of the new relationship in clients' eyes, joint IC development serves to bring colleagues together from across companies, now collaboratively focused on a common goal.

S4.4.3 Tools and Techniques

Our IMO uses standard program/project management tools and techniques, such as:

- **The aforementioned IMS:** describes in detail (often multi-thousand lines) the work activities, how they are linked, their schedule, and who is performing them.
- **Risks, Assumptions, Issues, Decisions, Dependencies (RAIDD) log**: we've learned the hard way that previous decisions and their context can be lost, especially as an IMO transitions operations to full-time market-facing leaders, without adequate documentation. Also, as stated previously, issues and risks don't "age" well if not addressed!
- **Dashboard**: we update our Balanced Scorecard performance measures, with supporting context, each month. To encourage transparency, the IMO and acquired entity jointly build the dashboard, which we brief to our Steering Committee.
- **Meetings**: we conduct recurring integration meetings, focused capture meetings, and program management reviews to status the new operation's business performance, encourage dialogue, and provide a forum for collaboration and understanding how each company approaches problems. We take great pains to avoid "death by meetings"—meetings are purposeful, prioritized, and outcome-oriented or they disappear!
- At the conclusion of each integration, we conduct a *lesson-learned exercise* where all stakeholders above provide input on what went well and not, with recommendations on how we can improve in the future.
- **Integration Risk Diagnostic**: Based on our internal experience, we've created an integration risk diagnostic that scores critical integration risk areas. When compiled, these individual risks present an overall score and snapshot that has proven accurate in predicting risk— "low," "medium," and "high" of a successful integration. We update the diagnostic—the risks, their weighting, etc.—after each integration.

S4.4.4 Conclusion

For companies like Booz Allen, M&A can be a key lever for strategic growth and overall business transformation. However, M&A presents inherent risk as two companies are brought together. Having a strong PMI approach is critical to improving the likelihood of success—meeting deal aspirations—but success is far from guaranteed. Booz Allen's PMI approach uses "hard" (e.g., IMS, RAIDD log, Balanced Scorecard performance measures, and WBS) and "soft" (e.g., culture exchanges and two-way integration) practices to bring together—and to bring the best out of—new sets of colleagues.

Tip

The Booz Allen Hamilton secret sauce has key ingredients, like living the purpose, unleashing courage, empowering, prioritization, transparency, and radical candor.

S4.5 Innovation Excellence

The next and last excellence enabler is covered in our recent article, Innovation using Skunk Works and project management, Kerzner and Zeitoun (2024).

Today, regardless of what periodicals or books you read that discuss project management practices, you will most assuredly find information discussing Agile and Scrum. What most people do not realize is that several of the principles of Agile and Scrum are more than 80 years old, having been used by Lockheed during the 1940s when it created the famous "Skunk Works" dedicated to radical innovation. Most people may have heard of "Skunk Works" but do not understand the impact it had on project management practices years ago and the impact it is still having in many companies worldwide.

S4.5.1 The Need for an Innovation Unit

One of the main drivers of a company's competitive advantage is innovation. Unfortunately, there are several types of innovation, and each type comes with advantages and disadvantages that may affect certain functional units. Let's consider just incremental (or a continuous small improvement) innovation and radical innovation.

The selection of the type of innovation can be impacted by the personal desires of the people who must make the decision and is often based upon how they feel about the status quo versus the future. Some companies are fearful of the radical innovations from Skunk Works because of the risks of accepting new businesses. Examples would include Xerox and personal computers, as well as Kodak's failure in digital photography. These companies focused mainly on the expansion of core businesses.

Many executives prefer to promote short-term results such as in established businesses that generate sales, profits, and current executive compensation and reward packages rather than radical innovation where the results may not be known for years and are accompanied by financial uncertainties. When executives resist major changes, they then assign their brightest and most talented people to short-term results and commercialization of new ideas may suffer.

Functional units can also resist new technologies if there is a fear of being removed from their comfort zones. Changes in technology can be accompanied by additional costs such as purchasing new equipment and facilities, hiring new workers, developing new procedures, retraining expenses, and new marketing and sales requirements.

The resistance to changes in technology can trigger competition between functional departments such as R&D and manufacturing. The unfortunate result in some companies is when manufacturing resists radical innovation practices that could favorably impact the organization's future. To overcome the resistance problem, companies created a so-called Skunk Works unit for radical innovation where the unit is isolated from the parent organization. The traditional R&D organization would then be responsible for continuous improvement projects, and the Skunks Work unit would manage radical innovation activities.

S4.5.2 The Birth of "Skunk Works"

During the early years of World War II, the United States and its allies realized that their fighter planes were no match for Germany's new jet fighters. The U.S. War Department asked Lockheed for help in 1943. Lockheed created a special unit entitled Lockheed's Advanced Development Program. Later, the name of the program was given the pseudonym "Skunk Works."

The term "Skunk Works" came from Al Capp's hillbilly comic strip Li'l Abner, which was popular in the 1940s and 1950s. The original term in the comic strip, "Skonk Works," was a dilapidated factory that generated strange odors and was located on the remote outskirts of Dogpatch. The Lockheed unit began using the term "Skunk Works," thanks to an engineer in the original team who was a fan of the comic strip.

The special unit was headed up by Kelly Johnson, Lockheed's 33-year-old chief engineer, who ran the unit for almost 45 years. His nickname at Lockheed was "Engineer of the Century." The intent was to create a special team composed of a small group of some of Lockheed's most talented employees, hand-picked by Kelly, to work on secret projects that required innovation.

To help maintain secrecy and avoid distractions, the team was allowed to work autonomously at a secret location away from distractions that could come from Lockheed's main operations. Kelly was provided with a limited budget to support the effort and aggressive schedules.

From a project management perspective, Kelly was the program manager responsible for all of the secret projects within Skunk Works. However, as chief engineer at Lockheed, he also had to share his time each day with ongoing activities at the main operations unit that were not part of Skunk Works. Kelly was highly successful in his tenure of running Skunk Works for 45 years.

During his tenure, Kelly developed 14 "Rules" for all Skunk Works projects which were directly related to most project and program management practices requiring innovation. The "Rules," most of which still apply today, will be discussed later in this case study. Ben Rich, who eventually replaced Kelly, also promoted the 14 "Rules." The result has been an ongoing record of innovations at Lockheed for more than 70 years.

S4.5.3 Challenges with "Skunk Works" Growth

Companies that need innovation for growth and survival have heard of Skunk Works and recognize the application for running secret projects using the best people available. Skunk Works thrives on self-driven teams that focus on making breakthrough innovations in a reasonably short time frame. The selection of the researchers for Skunk Works is critical. They must enjoy the research and experimentation needed in dealing with the risks and uncertainties that could lead to major innovation breakthroughs. They must also possess a passion for teamwork and cooperation with colleagues at Skunk Works. Knowledge of project management is most certainly helpful.

Skunk Works shows the entire company where technology may be heading, and this is accomplished without spending a great deal of money. The result is most often better decision-making on opportunities involving creativity. As stated by May Matthew,[4]

> **"High-quality designs in a short time frame with limited resources are the hallmarks of a Skunk Works project."**

The Skunk Works approach has been used successfully by numerous companies. Steve Jobs used it to launch the Macintosh computer at Apple, as well as the iPhone and iPad. Ford Motor Company used Skunk Works to rapidly integrate technology into useful automotive features. Disney created an entire division entitled "Imagineering" (i.e., IMAGination and engINEERING) to function as an R&D laboratory to bring stories to life. The division is remotely located from Disney's headquarters and functions as the creative unit that designs and builds all Disney theme parks, resorts, cruise ships, games, publishing, movies and cartoons for TV and theater screens, and product development businesses. IBM used Skunk Works to create personal computers. Microsoft also used Skunk Works to develop

---

4 May, Matthew E., Skunk Works: How Breaking Away Fuels Breakthroughs, *Rotman Management*. Spring 2013, p52-56.

computers and tablets. HP created pocket calculators, laser printers, and 3D printers using Skunk Works.

Other well-known companies using Skunk Works included Google, DuPont, Boeing, GenCorp, Siemens, Philips, Intel, LEGO, and Xerox. The management guru, Tom Peters, co-authored a book entitled "A Passion for Excellence" in which Skunk Works was highly praised as a means for innovation, competitiveness, and growth.

Some companies focus on part-time innovation. Google's "20% time" policy allows employees to spend one day a week working on projects, even though they may have other responsibilities. The results were Google News and Gmail. To promote this policy, Google demands that at least 30% of each division's revenue come from products introduced within the past four years. This impacts employees' bonuses and salaries. A similar policy exists at 3M. Employees are allowed to spend 15 minutes each day thinking up new products for 3M, and at least 25% of the division's revenue must come from products introduced within the past five years.

Most of today's companies have recognized the need for innovation, creativity, design thinking practices, and advances in technology. Yet many of the companies have not given consideration to Skunk Works as a possible means for growth because of their interpretation and fear of the accompanying challenges. Lockheed was able to overcome the challenges, but even with their success, they admitted there would be limitations for others. Ben Rich, who served as Vice President and General Manager at Lockheed's Skunk Works, discussed the challenges some companies will face based on his experience with government projects[5]:

> **"I seriously doubt that most of these companies will successfully implement the Skunk Works' management style, however. In many, if not most, cases, it's the wrong thing to do. There are too many outside factors that hinder implementation of the Skunk Works' philosophy, not the least of which is the number of requirements imposed by the United States government."**

Even today, Skunk Works is considered by many as restricted mainly to large and expensive high-technology projects specifically designed for aerospace and defense units of the U.S. Government. This is certainly not true. Lockheed's success has been with small as well as large projects requiring creativity.

Results have shown that the successful marriage between Skunk Works and project management practices can lead to innovation efficiency. Unfortunately,

5 Rich, Ben R., The Skunk Works Management Style: It's No Secret, Vital Speeches of the Day. 11/15/88. Vol. 55 Issue 3, p87–93.

creating an innovative product, even quickly, is no guarantee that there will exist a market demand for the product. There must exist a business need for creating a Skunk Works unit. Some units fail to develop a strategy for commercializing the innovation outcomes. Peter Gwynne identified challenges that Xerox faced and how they addressed the challenges:[6]

> **"To be successful with them (Skunk Works), they have to be business oriented–that is, they must create successful businesses rather than successful products. So, Xerox is now taking a new approach to Skunk Works: Starting up projects as small businesses with their own P&L responsibility and marketing personnel, rather than internal groups that have to rely on the corporation for those activities and people."**

In most of the Fortune 100 companies, project management is more than just another career path. It is seen as a strategic competency necessary for the growth of the organization. As a PM, you are now seen as managing part of a business rather than just a project. You are expected to make business decisions as well as traditional project decisions. Most projects today that focus on innovation outcomes include a life cycle phase entitled commercialization.

Unlike traditional product improvement R&D that might focus on finding higher quality raw materials or cheaper ways to manufacture the products, Skunk Works has a significant business component that includes prototype development, reducing time-to-market, developing their own channels of distribution, and selling the products directly to the customers.

Developing innovative products does not maximize business benefits to a company unless the innovation team is allowed to make the necessary time-to-market commercialization decisions to take advantage of opportunities. As stated by Single and Spurgeon in a discussion about Ford's Skunk Works,[7]

> **"All automotive companies are working hard, with considerable success, to reduce the time from concept to customer for vehicles. It is necessary to do the same thing for innovative features. Companies that learn how to do this will certainly have a competitive edge. A well-designed Skunk Works is an eminently practical way of accelerating the implementation process."**

---

6 Gwynne, Peter. Skunk Works – 1990s Style, *Research Technology Management*. Jul/Aug1997, Vol. 40 Issue 4, p18.

7 Single, Arthur W., Spurgeon, William M., Creating and commercializing innovation inside a Skunk Works, *Research Technology Management*, 08956308, Jan/Feb1996, Vol. 39, Issue 1.

Another challenge with Skunk Works is the culture that is created. Implementing Skunk Works has forced senior management to rethink the issues of allowing multiple cultures to exist concurrently. For years, companies allowed each project to have its own culture knowing that the projects would eventually come to an end. As companies began realizing that project teams must make business decisions, a single corporate culture was created in many companies that supported all types of projects and traditional business practices. Skunk Works cultures in most companies appear to be business-oriented, but they must also be product-innovative-oriented. As such, most people view Skunk Works as countercultural to protect the team from possible disagreements with the corporate culture.

Cultural differences can lead to misalignment issues. Misalignment in the relationship between the primary organization and Skunk Works. The greater the misalignment, the greater the chance that some good opportunities might be discarded, and other ideas might be promoted that are too risky and not in line with corporate goals and objectives.

Skunk Works thrives when team members can use unconventional approaches to problem-solving and decision-making, regardless of the size of the projects or programs. This often scares some executives who are afraid that implementation might cause senior management to lose control of the company by eliminating bureaucratic red tape needed for product checks and balances and reducing the time needed for approvals and decision-making. Skunk Works have minimal managerial constraints.

Project management has matured significantly since Ben Rich delivered his speech (see footnote 2) more than 30 years ago. The benefits of using project management and the accompanying best practices appear in numerous publications. Yet there still exists inherent fear of the Skunk Works approach in some organizations.

In many companies, project and program governance still resides at senior management levels because executives do not trust project teams to make certain decisions that were traditionally reserved for senior management. Senior management also preferred to monopolize customer communications. This is contrary to what Lockheed did by allowing project teams the autonomy to develop close working relationships with customers and stakeholders.

There is also the fear among companies that might have government contracts that they will be expected to allow heavy involvement by government stakeholders and be burdened with an excessive number of legal policies and procedures that must be followed. Companies may fear that this may trickle down to non-government contracts as well. This is especially true with the growth of AI applications and concern over possible product liability lawsuits. Companies may not realize that many of these outside factors that existed previously had been restricted only to government projects and programs.

Another critical issue is the size of the company. As stated by Ben Rich,[8]

> **"I don't think a 'Skunk Works' would be feasible if it couldn't rely on the resources of a larger entity. It needs a pool of facilities, tools and human beings who can be drawn upon for a particular project and then returned to the parent firm when the task is done."**

Company size today is no longer an issue for successful project management to exist but may impact the decision to implement Skunk Works. Even the smallest of companies can implement successful project management practices.

Some companies have used Skunk Works to respond to a customer's request for proposal (RFP). The response might include a prototype that underwent inspection and testing. If the company's bid is not accepted, the unit is dissolved, and people return to their previous organization. If the bid is accepted, the unit begins commercialization.

In some extreme situations, a company might establish multiple Skunk Works to bid on the same RFP. In this situation, the units are also in competition with each other to win the opportunity to submit their bid using their designs and therefore do not communicate with each other or share information. These units may include contracted labor.

Management would select the best innovative approach from one of the units for their bid. The other units are then dissolved. There are several risks associated with doing this. Other than cost, the use of contracted labor can create issues with secrecy and control and ownership of intellectual property.

Perhaps the most important lesson learned from Skunk Works is the need to develop a corporate project management culture that can bring out the best in people, and this often requires an unconventional approach to project leadership where team members are effectively engaged throughout the life of the project or program. If this is done correctly, creativity will follow and lead to success. The challenges and issues can be overcome.

S4.5.4 Kelly's 14 Rules and Practices at Skunk Works[9]

Kelly's 14 rules were specifically designed for Lockheed's Skunk Works. Today, most of these rules still apply and may be highly beneficial to all companies, especially those needing innovation and creativity. The rules are designed around project and program management practices that have been highly successful at Lockheed for more than 70 years. The rules will be discussed from a project management perspective.

---

8 See footnote 2.

9 The rules can be found at lockheedmartin.com/us/aeronautics/skunkworks/14rules.html.

Rule #1: The Skunk Works® manager must be delegated practically complete control of his program in all aspects. He should report to a division president or higher. Innovation decisions that must follow the chain of command and obtain everyone's input and approval can be time-consuming and slow down the decision-making process. Project governance on many types of projects works best with individual rather than committee sponsorship, and that individual should reside near the top of the organizational chart. Single-person governance can also eliminate having to work with often hidden agendas of many managers who wish to participate in decisions involving innovation for personal reasons rather than for what is in the company's best interest.

Rule #2. Strong but small project offices must be provided both by the military and industry. Not all projects and programs can be managed by a single person. Some projects require the creation of a project office composed of assistant PMs. Clients like the U.S. Government often demand that a government project office also exist on site as a means of tracking performance on some high-visibility government programs. When this occurs, contractors often assign the same number of people in their project office as the customer would have in their project office to provide one-on-one coverage and communications. Large project offices increase overhead, increase communication channels, slow down decision-making, and increase the project's overhead costs.

Rule #3. The number of people having any connection with the project must be restricted in an almost vicious manner. Use a small number of good people (10% to 25% compared to the so-called normal systems). Strategic projects, especially those that require innovation and creativity, have a much greater need for problem-solving and decision-making practices. The larger the number of people connected to the project, the greater the number of channels of communication that must exist. This can take a great amount of time and increase a project's budget. By restricting the number of people connected to the project, decisions can be made in hours or days rather than weeks or months. Action items are more quickly resolved.

Rule #4. A very simple drawing and drawing release system with great flexibility for making changes must be provided. This rule was created before we had computer-aided design and computer-aided manufacturing (CAD-CAM). The intent, which still exists on projects requiring drawings, is to make it easy for changes to be made and approved.

Rule #5. There must be a minimum number of reports required, but important work must be recorded thoroughly. We have all written reports that are never completely read. Report preparation is costly and involves writing, typing, proofing, editing, approvals, reproduction, security classification if necessary, and even disposal. The cost, fully burdened for everyone involved, could exceed $2000 per page. Reports should be minimized but accurate and include all the critical information.

Rule #6. There must be a monthly cost review covering not only what has been spent and committed but also projected costs to the conclusion of the program. This rule has become standard as a part of all project monitoring and control reporting systems. Reporting today includes the estimate at completion (EAC) as well as actual and budgeted costs.

Rule #7. The contractor must be delegated and must assume more than normal responsibility to get good vendor bids for subcontract on the project. Commercial bid procedures are very often better than military ones. Even in today's environment, government customers in some countries still dictate to contractors how to evaluate suppliers and which suppliers they can use. In one country, the local government forced contractors to select suppliers only from within the country and to give favoritism to suppliers in cities that had the greatest unemployment rates. Topics such as cost, quality, and lead times were of secondary importance.

Rule #8. The inspection system as currently used by the Skunk Works, which has been approved by both the Air Force and Navy, meets the intent of existing military requirements and should be used on new projects. Push more basic inspection responsibility back to subcontractors and vendors. Don't duplicate so much inspection. As discussed in Rule #2, customers and government agencies often establish project offices on the contractor's site. This can lead to duplication of inspection practices and can force contractors to establish multiple inspection processes based on customer requirements.

Rule #9. The contractor must be delegated the authority to test his final product in flight. He/she can and must test it in the initial stages. If he doesn't, he rapidly loses his competency to design other vehicles. Allowing government and military personnel to have the responsibility for product testing can create issues if the personnel are rotated to different assignments during the project and new people appear with a different interpretation of how good the product works. Product testing is the responsibility of the company that must design and manufacture the product. Testing should be done throughout the life cycle of the project to minimize the risks of downstream product liability lawsuits.

Rule #10. The specifications applying to the hardware must be agreed to well in advance of contracting. The Skunk Works practice of having a specification section stating clearly which important military specification items will not knowingly be complied with and reasons therefore is highly recommended. If appropriate, all contracts and even SOWs should have a specification section. Project teams must clearly understand specification requirements before the final contract price is agreed to.

Rule #11. Funding a program must be timely so that the contractor doesn't have to keep running to the bank to support government projects. Customers often underfund contracts just to get the work started. Contractors often grossly underbid the initial contract and then either ask for additional funding or try to push through scope changes. In either case, both the contractor and customer must have a clear understanding of the cost of the project and the available funding to match the cost.

Rule #12. There must be mutual trust between the military project organization and the contractor to achieve ***close cooperation and liaison on a day-to-day basis. This cuts down misunderstanding and correspondence to an absolute minimum.*** Trust has become perhaps the most important word in project management. One of the reasons why customers establish a project office at the contractor's location is to minimize paperwork and reduce misunderstandings. Customer communication in the past was at a minimum because contractors believed that customers and stakeholders did not understand project and program management and would meddle in the daily operations of the projects. Today, customers and stakeholders possess project management knowledge, and their help and advice are welcomed.

Rule #13. Access by outsiders to the project and its personnel must be strictly controlled by appropriate security measures. Outsiders often go to extreme measures to find out what projects your company might be working on to bring this knowledge back to their organization. One company even went so far as to find out the salary of certain people working on secret innovation projects and then offered them a larger salary to change companies.

Rule #14. Because only a few people will be used in engineering and most other areas, ways must be provided to reward good performance by pay, not based on the number of personnel supervised. People should be paid and rewarded for their accomplishments rather than the size of their empire.

S4.5.5 Project Management Practices Within Skunk Works

The ability of the team to collaborate with each other, respect each other's opinions, and a willingness to participate in group decision-making are mandatory for increasing the chances of Skunk Work's success. These are some of the reasons why the participants in Skunk Works are most often hand-picked by the leader. Gaining the benefits of a successful Skunk Works may require organizations to rethink how project management should be implemented within the unit.

Flexibility and the use of techniques such as Agile or Scrum are beneficial. Projects that have a heavy focus on innovation often follow different practices than traditional projects that begin with well-defined requirements that may remain fixed over the life cycle of the project. Project management practices within the Skunk Works should include the following:

- Project planning may need to be structured around short time periods, such as sprints.

- At the end of each period, continuous improvement decisions must be made based on experimentation, inspection, observation, and experience.
- Team members must recognize the need for continuous feedback and that project success is based upon iterative development.
- Teamwork should be seen as the driver for success.
- Project team members must respect each other and the recommendations and decisions others might make.
- Collaboration with team members and stakeholders is more important than relying upon tools and processes.
- Project documentation should be minimized if possible.
- Project teams must be prepared to make business and product commercialization decisions.
- Project teams must be willing to be removed from their comfort zones and work on tough problems.
- Safeguarding intellectual property is critical.
- Business metrics that focus on business goals and objectives should be used along with traditional project metrics.

Many of the abovementioned bullets are the characteristics of Agile and Scrum project management practices. There are certainly other factors that could be included.

S4.5.6 Conclusion

The need for innovation and new products will most certainly increase. In the future, more companies are expected to consider Skunk Works as a possible solution to corporate growth. Combining the abovementioned bullets with Kelly's 14 rules provides us with a glimpse of how project management practices take place in Skunk Works. Effective project management practices can lead to innovation and commercialization success. But it will be challenging for some companies.

> **Tip**
> A well-designed Skunk Works is an eminently practical way of accelerating the implementation process. This requires a culture of experimenting and engaging.

Index

Creating Experience-Driven Organizational Culture: How to Drive Transformative Change with Project and Portfolio Management, First Edition. Al Zeitoun.